The Boundary Flux Handbook

The Boundary Flux Handbook

A Comprehensive Database
of Critical and Threshold Flux
Values for Membrane Practitioners

Marco Stoller
La Sapienza University of Rome
Department of Chemical Materials Environmental Engineering
Rome
Italy

Javier Miguel Ochando-Pulido
University of Granada
Department of Chemical Engineering
Calle Real de Cartuja
Spain

ELSEVIER

AMSTERDAM BOSTON HEIDELBERG LONDON NEW YORK OXFORD
PARIS SAN DIEGO SAN FRANCISCO SINGAPORE SYDNEY TOKYO

Elsevier
Radarweg 29, PO Box 211, 1000 AE Amsterdam, The Netherlands
The Boulevard, Langford Lane, Kidlington, Oxford OX5 1GB, UK
225 Wyman Street, Waltham, MA 02451, USA

ISBN: 978-0-12-801589-6

British Library Cataloguing-in-Publication Data
A catalogue record for this book is available from the British Library

Library of Congress Cataloging-in-Publication Data
A catalog record for this book is available from the Library of Congress

For information on all Elsevier publications visit our
web site at http://store.elsevier.com/

Working together
to grow libraries in
developing countries

www.elsevier.com • www.bookaid.org

Contents

About the Editors

Marco Stoller, born in Bonn, Germany, on December 16, 1974, and of German-Italian nationality, is an active researcher in membrane technologies, in particular in the area of membrane fouling. His research work started in 2000, during his Ph.D. thesis entitled "Development of a purification process of olive mill wastewater by membranes," and has been carried out continuously up to the present day. He is Assistant Professor in the Department of Chemical Materials Environmental Engineering at the University of Rome "La Sapienza." He is author of more than 35 papers in international peer-reviewed scientific journals, more than 70 contributions to conference proceedings, and more than 25 oral presentations at international congresses. In 2009, he was invited to give a lecture to the younger researchers at the Network Young Membrains 2009 event, held in Mezé just before Euromembrane 2009. In 2013, he was awarded the "Certificate for Excellent Presentation" for the research work "Fouling inhibition and control by threshold flux estimation in the treatment of olive vegetation wastewater by membranes-in-series process" at the 1st International Conference on Desalination using Membrane Technology, promoted by Elsevier.

Dr Stoller has taken part in six European projects, three of them involving the application of membrane processes, and has contributed to the dissemination of the fundamentals and applications of membrane processes in some master courses devoted to professional qualifications and in some conferences in the wider community. He is an active member of the Italian Chemical Engineer Association since 2001, member of EMS since 2004 and member of EFCE, Section on Membrane Engineering, since 2007.

Javier Miguel Ochando-Pulido, born in Jaén, Spain, is a young but active researcher in the membrane field, especially in reverse osmosis, nanofiltration and ultrafiltration technologies, and fouling issues. He completed his undergraduate studies in the University of Granada (Spain), and afterward received a state grant from the Science and Innovation Spanish Ministry to do his Ph.D. thesis on olive mill effluent reclamation in the Chemical Engineering Department of the University of Granada.

During his postgraduate studies, he also received a state grant from the Science and Innovation Spanish Ministry to do a residence period in the Department of Chemical Materials Environmental Engineering at the University of Rome "La Sapienza," under the direction of Prof. Angelo Chianese and Dr Marco Stoller. During this time, he improved his knowledge of advanced oxidation processes and membrane technologies for industrial effluent treatment.

In December 2012 he presented his Ph.D. thesis, entitled "Application of membrane technology for the depuration of olive mill wastewater." In 2013, he was awarded the "Certificate for Excellent Presentation" for the research work "Fouling inhibition and control by threshold flux estimation in the treatment of olive vegetation wastewater by membranes-in-series process" at the 1st International Conference on Desalination using Membrane Technology, promoted by Elsevier. He continues his postdoctoral research in the Chemical Engineering Department of the University of Granada, where he works as a lecturer. He is author of 24 papers published in international peer-reviewed scientific journals, more than 30 contributions to conference proceedings, and several oral presentations at international congresses.

Preface

Many times membrane practitioners need to know, as quickly as possible, how membranes will perform, given a specific feedstock and separation target, when integrated into an existing process or when designed from scratch for new processes. Unfortunately, many times it is not possible to provide a quick and reliable answer.

The main problem is to estimate a priori membrane fouling issues: how they will affect the longevity of the membrane modules and, as a consequence, impact the overall economic balance of the process. The longevity of the membrane modules is one of the main factors, together with their selectivity and productivity, in properly designing membrane applications.

In response to this need, one of the authors, Marco Stoller, started to develop a personal database based on all the case studies investigated during his career. The database was useful for both research and counseling purposes: for research, it was useful to define fouling problems before starting the experimental setup and campaign in order to have an idea of the experiment's run times and to save the membrane from premature blocking. In the case of counseling, with the complementary aid of proper tools, the database was successfully used for quick and reliable estimation of membrane performance in specific systems.

The problem arose as soon as the database was not extended enough to have useful information on every possible system. A personal database turned out to be too limited to cover all cases faced during work as a membrane practitioner.

The point at which it was necessary to extend the database was therefore reached.

During this period of time, a Ph.D. student named Javier Miguel Ochando-Pulido, from the University of Granada in Spain, was visiting the Department of Chemical Engineering to study membrane processes and membrane fouling. In the following years he managed to efficiently apply the critical and threshold flux concepts during his research. He expressed the same need to have an extended database available to support his work.

The decision to write this handbook was, therefore, finally made at the beginning of 2013.

The main core of this handbook—that is, the database of critical and threshold (hereafter merged as boundary) flux values—is based on published data. Sometimes the values were measured and given directly as a main result, but many times a successive evaluation by the authors was required to extract the useful information on fouling from the information included in the papers.

Personal gratitude from the authors to all contributors, as listed in the references section of this book, is hereby acknowledged and expressed.

The authors also want to thank all the people supporting their studies and research:

- The Chemical Plants group at the Department of Chemical Materials Environmental Engineering at the University of Rome "La Sapienza," in particular Prof. Angelo Chianese, full professor in the department, who promoted and supported the critical and threshold flux research.
- the Chemical and Biochemical Technology Research Group of the Chemical Engineering Department at the University of Granada, in particular Prof. Dr Antonio Martinez-Ferez, associate professor in the department, who assisted in and facilitated the membrane technologies research in the laboratories.
- Our respective families, who supported the authors during their studies: Dr Hermann Stoller, Latina Bolzoni, Luisa Conte, and Alice Stoller; and María Rita Pulido Bosch, Miguel Ochando Cruz, Eduardo Ochando Pulido, Antonio Pulido Peramos, and Encarnación Cruz Miranda.

This handbook will not represent an endpoint to research. It represents the actual state of the art in using commercially available membranes inhibiting fouling issues (and, thus, a starting point to improvement of their performance through further research) and a useful tool to help membrane practitioners design membrane processes.

Rome, June 24, 2014

Dr Ing. Marco Stoller
on behalf of the authors

Index of the Database

THE BOUNDARY FLUX DATABASE

Chapter 1

Introduction

In the last years the use of membrane technologies as stand-alone, integrated, or substitutive processes has been sensibly increasing, and the related market is becoming more active. The reason lies in the many advantages this technology may offer, from both a technical and an economic point of view, if compared to conventional ones. The availability of new membrane materials, membrane design, membrane module concepts, and general know-how promotes credibility in investors. Unfortunately, one main drawback remains membrane fouling. In past decades, this problem has perceptibly affected the reliability of the technology, and despite the development, understanding, and validation of the fouling phenomena theories and their description in membrane processes in order to overcome the knowledge gap, this feeling still persists. The lack of development of reliable tools to inhibit fouling represents a main constraint holding membrane technologies away from a definitive maturation.

Currently, proper membrane process design of systems affected by fouling issues can be a difficult task to accomplish. Plant designers target the project variables concerning productivity and selectivity, but may not be in the position to estimate correctly the membrane longevity. The estimation of this parameter, which mainly affects the economics of the membrane process, is based on experience and supplier's data, and this may not be sufficient as a guarantee to investors who demand precise balance calculations.

In the presence of fouling, the longevity of the membrane modules and the constancy of the permeate fluxes as a function of time are equally as important as the project targets concerning productivity and selectivity [1]. Fouling may lead to an irreversible reduction of the permeate flux rate as a function of time. The main consequence is a shorter lifetime of the membranes and an increase of the operational costs given by the substitution of the clogged modules. In order to overcome this uncertainty, designers tend to overdesign the membrane plant [2]. In most cases, the overdesign is performed by using high safety margins or by past experience of the designer, starting only from the knowledge of the permeate project value, without considering in detail the entity and nature of fouling. This approach was considered to be successful in order to guarantee the process performances, and investors feel comfortable using a process characterized by many advantages at increased costs.

The Boundary Flux Handbook. http://dx.doi.org/10.1016/B978-0-12-801589-6.00001-2

In more recent years, as investments are becoming more difficult, and investors pay more attention to expenses, the overdesign must be limited in order to target a profitable operation of the membrane plant. Designers are forced once again to take the risk to underdesign the membrane plants in order to remain competitive, as a consequence risking again the reliability of the technology. Unfortunately, the diffusion of this problem still persists.

The reports on membrane process failures are well hidden, and data is of confidential nature. Nevertheless, investors carefully review the actual state of the art on the technologies, and membrane fouling remains one main problem for this technology. The impression is that a general lack of knowledge exists, and membrane practitioners have difficulty quickly answering and precisely estimating membrane fouling issues when such information is requested and no experience exists beforehand.

In this context, it is not mandatory to define and describe with scientific precision the phenomena of membrane fouling. The causes may be of interest to the process designer, but not to the investor, who will ask about the consequences. In case of membrane fouling, when triggered, the consequences are in most cases negative.

One approach to answer the investor's need to trust membrane technology is to guarantee that fouling will be strongly inhibited or avoided. Since the description of the fouling phenomena may be difficult and complex, engineers may avoid the consequences altogether by limiting the causes. This choice sensibly affects the process design and requires reliable tools to evaluate the correct operating conditions.

For liquid—liquid separation processes, Field et al. introduced the concept of critical flux for microfiltration, stating that there is a permeate flux below which fouling is not promptly observed [3]. In subsequent years, researchers identified critical flux values for ultrafiltration ("UF") and nanofiltration ("NF") membranes, as well [4]. The concept is suitable to the need to avoid operating conditions triggering fouling, and as a consequence, to avoid fouling at all. Nowadays, the critical flux concept is well accepted by both scientists and engineers as a powerful membrane process optimization tool as long as critical fluxes apply [5]. Some examples of existing over- and underdesigned membrane plants are reported in Table 1.1.

Unfortunately, most systems treating complex solutions and real wastewater streams do not strictly follow the critical flux theory. Le Clech et al. noticed that on certain systems, operations below the critical flux may not be sufficient to have zero fouling rates [10]. In addition to this, the measurement of critical fluxes was often not possible, and in order to overcome this problem, the identification of "apparent" critical points was used. To overcome this limitation in the definition of the critical flux, in a recent paper, Field and Pearce introduced for the first time the concept of the threshold flux [11]. To summarizing this concept briefly, the threshold flux is the flux that divides a low fouling region, characterized by a nearly constant rate of

TABLE 1.1 Average Overdesign of Existing Membrane Plants

Feedstock Stream	Membrane	Overdesign (%)	Plant Location	Reference
Brackish water for potable water preparation	UF	319%[a]	Appleton, US	[6]
Wastewater discharge with high salt load in Rhein river	NF	230%	Germany	[7]
Sea water for desalination	RO	246%[a]	Al-Birk, S. Arabia	[8]
Domestic wastewater for treatment by MBR	UF	240%[a]	Private household	[9]

[a]Estimated as observed underdesign during operation.

fouling, from a high fouling region, where flux-dependent high fouling rates can be observed.

The extension of the critical flux theory to the threshold flux theory is important because the benefits were extended to those systems treating complex feedstocks. Nowadays, the threshold flux concept appears to be a powerful membrane process optimization tool that finds positive feedback and successful validation.

Unfortunately, the identification of a trustable tool may not be sufficient if the application is difficult. The main drawback of the application of critical and threshold fluxes on systems is the fact that the determination of the values cannot be theoretically predicted, but can only be experimentally determined by time-consuming experiments at a certain moment in time. Moreover, different threshold flux values can be measured on the same system, depending on various factors, such as hydrodynamics, temperature, feed stream composition, and membrane surface characteristics [5,12−14]. Feed stream composition and operating time are the main factors responsible for ever-changing threshold flux values, and this is especially true for treatment of real wastewater streams by membranes, since the entering feedstock quality is not constant during the year. An additional factor is exemplified in batch membrane processes, which are mainly used when the availability of the feed stream is limited or discontinuous, and the concentrate stream is recycled back to the feed tank, thus irremediably leading to sensible feedstock changes during operation.

As a consequence, threshold flux values never remain constant. This represents a major difficulty in fine-tuning optimal operating conditions and in applying this tool successfully to membrane process design purposes. Nevertheless, critical and threshold flux theories were quickly used by researchers and membrane process designers in order to inhibit membrane fouling in many systems. As an illustration of the popularity of the use of these concepts, more than 6400 papers were published in international scientific journals between 2010 and 2014 [15]. The concepts are not able to describe and explain the fouling phenomena, but give advice on how to avoid it; and this may be fully sufficient in order to properly design membrane plants.

The problem of correctly defining ever-changing critical and threshold flux values leads to confusion among membrane practitioners. Even in research, among academics, confusion about the correct use of the theory exists. Some researchers have published comprehensive review papers about the correct use of the concepts [5,11]. Critical fluxes were defined in different types, such as strong form, weak form, and for irreversibility. Despite these efforts, the introduction of new critical flux definitions has increased the confusion and the erratic use among membrane practitioners, and the problem has become even worse with the introduction of the threshold flux concept. Since late 2011, in international scientific journals covering the treatment of wastewater streams, besides those exiting the olive oil, fruit juice, textile, and biotech (algae) industries, only 32 papers were published that used the term "threshold flux" correctly [15,16−23]. Many papers still rely on a wrong use of the older critical flux concept in those systems where threshold fluxes suit more precisely.

In order to overcome these problems, many designers use the sustainable flux as a project parameter. This flux is equal to the set point value of the control system of the membrane plant in order to achieve the productivity targets, even if fouling is formed at such a rate that it does not affect the output over long periods of time. Compared to the other definitions, the sustainable flux has a subjective characteristic, trying to fill the knowledge gap by the experience of the designer; as such, it cannot be considered a valid engineering tool for design purposes, although it was successfully adopted many times.

By way of an example, the authors of this work, both academic and chemical engineers, were not immune to this problem. The sustainable flux concept is difficult to consider as a possible parameter in order to develop design tools. In the past years, before the concept of the threshold flux was introduced, many papers on olive vegetation wastewater purification by membranes, mainly UF and NF, were published by the authors, always determining strangely behaving critical fluxes [24−28]. Irreversible fouling arises quickly on the membranes due to the high concentration of pollutants when wastewater is purified without any pretreatment, and different pretreatment processes influence to a variable extent the critical flux values [29−31]. Therefore, proper and optimally designed pretreatment processes on the given feedstock must be developed in order to maximize productivity and minimize fouling: henceforth this objective will be referred to as the concept of pretreatment tailoring of membrane processes. The authors observed in previous research work the change of fouling regime by using olive mill wastewater [18,28,29]. Despite that the applied optimization methods were based on modified critical flux measurements, before the threshold flux concept ever existed, successful fouling control was accomplished on this system and was justified in detail by means of the threshold flux in 2013 [18]. The developed tools were successfully validated, but they depended on the existence of many different concepts, thus, and as a consequence, their correct application may be result very difficult.

Field et al. suggested that the correct application of the threshold flux is de facto very near to the sustainable one [11]. Moreover, critical and threshold flux concepts share many common aspects that merge perfectly into a new concept, that is, the boundary flux. The introduction of the new boundary flux concept does not extend by addition of new theory or knowledge the critical and threshold flux concepts. Instead, it tries to simplify the use of these concepts in future works. Referring to one single concept will sensibly reduce the incorrect use of both the critical and threshold flux concepts and permit the development of simplified design tools for membrane engineers. Once these tools are widely available, by proper boundary flux knowledge or determination, membrane process sustainability may be reached or, at least, roughly estimated.

Chapter 2

The Boundary Flux

In this section the boundary flux will be introduced and widely discussed. The boundary flux shares all known aspects of the critical and threshold flux concepts, without adding something new. The advantage to using a general boundary flux concept lies in the impossibility of misinterpreting its value, and it eliminates the confusion among membrane practitioners in discriminating between different flux types. The boundary flux will be the guideline flux, and all successive considerations, from the optimal membrane performances down to the optimal membrane process design and plant construction, will be based on it.

The name was initially chosen randomly. After a friendly meeting with Robert Field during an International Congress in July 2014, and after hours of interesting debates on critical and threshold fluxes, both agreed that the boundary flux may be an appropriate name since fouling will trigger and thus strictly depending on what happen in the boundary layer over the membrane surface. Details will not be presented here in this handbook, since they are out of its scope, but findings will be available soon in the literature.

CRITICAL AND THRESHOLD FLUXES

Membrane fouling, expressed as a permeate flux reduction as a function of time given by some phenomena different than polarization and/or aging of the membrane, and thus as an increase of the total membrane resistance, can be subdivided into three main typologies:

1. A reversible fouling; this kind of fouling strictly follows the driving force amplitude, e.g., operating pressure values. As soon as the pressure over the membrane is reduced, this fouling is eliminated after a certain (short) period of time by the same quota.
2. A semi-reversible fouling; this kind of fouling accumulates over the membrane surface and cannot be easily eliminated. The only way to eliminate this kind of fouling is to stop the separation process and clean or wash the membranes, with water or aqueous solution of chemicals, respectively. Although this kind of fouling is almost eliminated after the cleaning/washing procedure, it represents a problem in the continuous process operation because it forces process shutdown at timed intervals.

The Boundary Flux Handbook. http://dx.doi.org/10.1016/B978-0-12-801589-6.00002-4

3. An irreversible fouling; once formed, this kind of fouling cannot be eliminated by any cleaning or washing procedure. It is the main cause of membrane failure concerning productivity.

In all cases, during operation of tangential cross-flow separation by membranes, all three fouling types will unavoidably appear and form to different extents. The existence of different fouling typologies affecting membranes was previously explained by Bacchin et al. and Oringer et al., and is based on the assumption of possible local conditions triggering different liquid/gel phases over the membrane and in the membrane pores due to the concentration profiles by polarization [32,33].

The distinction between semi-reversible and irreversible fouling is somewhat subjective, since there may be not-yet-considered washing or cleaning procedures that may eliminate some of the irreversible fouling more effectively, thus qualifying this segment as semi-reversible fouling. Therefore the difference in the nature of fouling depends strongly on the knowhow and the adopted cleaning protocol. On the other hand, other approaches to distinguishing irreversible fouling from the rest appear not possible, and must be related individually to each analyzed system.

For systems that exhibit only reversible fouling at low-pressure conditions, critical flux concepts apply best. In this case, although strictly not correct, membrane aging may be considered the only type of irreversible fouling affecting this system. Concerning the critical flux J_c, hereafter used in terms of critical flux for irreversibility, the following fitting equations apply [3]:

$$-dm/dt = 0; \ J_p(t) \leq J_c \tag{2.1}$$

$$-dm/dt = B(J_p(t) - J_c); \ J_p(t) > J_c \tag{2.2}$$

where m is the permeability of the membrane, B is a fouling fitting parameter, and $J_p(t)$ is the permeate flux at time t.

As soon as Eqn (2.2) applies, irreversible fouling starts to form. Semi-reversible fouling may be hardly observed in all the applied pressure ranges, and is therefore neglected. If semi-reversible fouling appears at low operating pressures, the system fits well with the threshold flux theory. Concerning the threshold flux J_{th}, the proposed equations by Field et al. are as follows [11]:

$$-dm/dt = a; \ J_p(t) \leq J_{th} \tag{2.3}$$

$$-dm/dt = a + b(J_p(t) - J_{th}); \ J_p(t) > J_{th} \tag{2.4}$$

where a and b are both fouling fitting parameters.

In case Eqn (2.3) is valid, both reversible and semi-reversible fouling are observed, whereas irreversible fouling may be neglected. As soon as Eqn (2.4) applies, irreversible fouling sensibly adds to the previous fouling phenomena.

It is interesting to note that the threshold flux equations are similar to the critical flux equations and differ only in the presence of the "a" parameter. In fact, if the case of "a" equal to zero is admitted, Eqns (2.3) and (2.4) may reduce to Eqns (2.1) and (2.2), respectively.

The parameter "a" value measures below threshold flux conditions the constant permeability loss rate of the membrane in time. If this value is equal to zero, no permeability will be lost in time and therefore no fouling is triggered. This is valid only below critical flux conditions, and therefore Eqn (2.3) includes Eqn (2.1) if a = [0, ∞). It appears that the parameter "a" is related to the semi-reversible fouling of the membrane system.

Above critical and threshold flux conditions, fouling behaves in a similar way by exponential permeability loss rates as a function of time. Again, if a = [0, ∞), Eqn (2.4) fits Eqn (2.2). The only difference between these systems is that in critical flux-characterized systems fouling is not affected by the continuous presence of a constant fouling permeability loss rate as in threshold flux-characterized systems. Besides this theoretical difference in the two systems, the authors want to point out that this aspect is of limited practical importance, since the exponential part of Eqns (2.2) and (2.4) will quickly overwhelm the linear contribution of the parameter "a" in Eqn (2.4).

Summarizing, both critical and threshold fluxes divide the operation of membranes into two regions: a lower one, where no or a small, constant amount of fouling is triggered; and a higher one, where fouling builds up very quickly.

MERGING BOTH CONCEPTS INTO ONE: THE BOUNDARY FLUX

The introduction of the new boundary flux concept does not extend by addition of new theory or knowledge the critical and threshold flux concepts. Rather, it tries to simplify the use of these concepts in future works. Referring to one single concept will reduce sensibly the incorrect use of both the critical and threshold flux concepts.

By introducing a new flux—that is, the boundary flux J_b—the previous equations may be written as:

$$-dm/dt = \alpha; \; J_p(t) \leq J_b \qquad (2.5)$$

$$-dm/dt = \alpha + \beta(J_p(t) - J_b); \; J_p(t) > J_b \qquad (2.6)$$

where:

- α, expressed in (L h^{-2} m^{-2} bar^{-1}), represents the constant permeability reduction rate suffered by the system and will be hereafter called the sub-boundary fouling rate index. α is a constant, valid for all flux values.
- β, expressed in (h^{-1} bar^{-1}), represents the fouling behavior in the exponential fouling regime of the system, and will be hereafter called the super-boundary fouling rate index. β appears to not be a constant, and changes with transmembrane pressure (TMP).

Although super-boundary flux conditions should be avoided to efficiently inhibit fouling problems, some words will be spent to describe these operating conditions, in particular the nature of the fitting parameter β. The interest regarding super-boundary flux conditions is here expressed with insight on existing processes affected by fouling, or by reducing the overdesign of plants, taking into account formation of a certain amount of irreversible fouling.

The method to determine the value of β is here reported by an example. In order to check the changes of β as a function of other parameters, a data set reported elsewhere was analyzed [28]. In Table 2.1 the relevant data set taken by using a nanofiltration (NF) membrane module is reported.

In Table 2.1, TMP is the applied transmembrane pressure to the membrane, TMP_b the measured transmembrane boundary pressure. The boundary flux $J_b(t)$ and the pure water permeability $w(t)$ were measured each time at the start $(t = 0)$ and at the end $(t = T)$ of the super-boundary operating conditions, holding this condition for a period of time equal to T minutes.

The chosen fitting equation for the data was:

$$\beta(TMP) = \zeta(TMP - TMP_b)^{\varepsilon} \qquad (2.7)$$

with ζ and ε being fitting parameters and TMP_b the boundary TMP, that is, the value of TMP where J_p is equal to J_b. In Figure 2.1, the fitting results of Eqns (2.6) and (2.7) are reported graphically.

The best fitting, by considering the maximum, minimum, and average mean square deviation of the results, was obtained by assuming a value of ε equal to unity; in other words, β appears to be proportional to the irreversible driving force; that is, the difference between the applied TMP and TMP_b. Concerning α and ζ, the best values were equal to $0.02 \text{ L h}^{-2} \text{ m}^{-2} \text{ bar}^{-1}$ and $0.0014 \text{ h}^{-1} \text{ bar}^{-2}$, respectively.

This result was expected, since it was observed many times that higher operating conditions lead to faster irreversible fouling. Moreover, with β being a function of TMP, Eqns (2.5) and (2.6) do not exhibit a discontinuity at the boundary point. The continuity of the two equations is therefore guaranteed, as it should be.

Therefore following equations hold at super-boundary flux conditions:

$$\beta(TMP) = \zeta(TMP - TMP_b) \qquad (2.8)$$

$$\lim[TMP \rightarrow TMP_b] \, \beta(TMP) = 0 \qquad (2.9)$$

PARAMETERS AFFECTING THE BOUNDARY FLUX VALUES

The method to measure the boundary flux is similar to the ones used to measure critical flux values, but needs a different approach in order to determine first the value of α and, successively, the value of β. Besides experimental data, the extended method requires the use of Eqns (2.5) and (2.6) to separate

TABLE 2.1 Super-Boundary Operating Conditions Data Set

ID	I	II	III	IV
TMP (bar)	30	30	20	20
T (min)	60	30	60	120
TMP_b (bar)	7	7	7	7
$J_b(0)$ $(L\ h^{-1}\ m^{-2})$	18.2	9.6	14.3	12.0
$J_b(T)$ $(L\ h^{-1}\ m^{-2})$	12.2	7.4	11.2	7.9
$w(0)$ $(L\ h^{-1}\ m^{-2}\ bar^{-1})$	8.42	8.25	8.00	8.13
$w(T)$ $(L\ h^{-1}\ m^{-2}\ bar^{-1})$	8.25	8.00	8.13	7.78

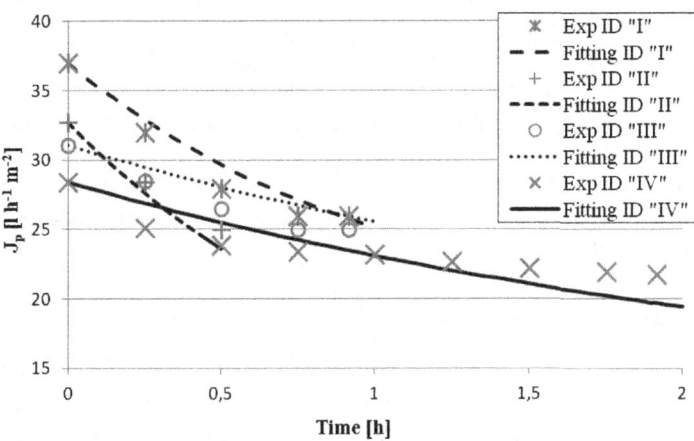

FIGURE 2.1   Fitting of the experimental data by Eqns (2.6) and (2.7).

the two different operating regimes, that is, at the end a region where irreversible fouling may be neglected from another region where irreversible fouling is formed. The validity of Eqn (2.5) excludes the validity of Eqn (2.6): as long as Eqn (2.5) holds, sub-boundary flux conditions are met. This approach needs a simple mathematical model to fit pressure cycle experimental data.

As soon as the dm/dt value diverges from constancy, the first point is the boundary one, and the relevant TMP value is equal to TMP_b.

The boundary flux values are sensibly influenced by the following parameters affecting the critical and threshold flux, described in detail below:

1. Hydrodynamics
2. Temperature
3. Membrane properties

4. Time
5. Feedstock characteristics
6. Washing and cleaning of membranes

Hydrodynamics

The increase of the feed flow (and, as a consequence, of the feed velocity and the Reynolds number) increases the corresponding J_b value, characterized by a logarithmic relationship. Reaching a certain limit, further feed flow increments insignificantly affect the boundary conditions. This behavior was observed to be valid for most systems, and follows a relationship such as:

$$J_b(Re) = J_b(Re \to \infty)\left(1 - e^{-A\ Re}\right) \qquad (2.10)$$

with A being a fitting parameter.

Although this relationship may enter the model, it is normally not used since the feed stream rate is controlled and is thus constant once defined and fixed. It is common practice to control the feed flow rate slightly higher than the start of the plateau value to assure best hydrodynamic operating conditions to the system.

In some systems, Eqn (2.10) appears to be not valid. Although the profile is very similar, the system may exhibit a maximum for a certain value of Re and then partially drop to a lower plateau for Re $\to \infty$. In this case, it appears mandatory to control the feed flow rate precisely at the maximum value or slightly above it.

Temperature

The increase of temperature leads normally to J_b value increments, due to the increase of permeability of the membrane. The limit is given by operating heating costs and membrane materials. In some isolated cases, intermediate temperature values may be the optimum, depending on the fouling phenomena. A general relationship may be as follows, and is limited by a minimum temperature value T_{min}, being normally the temperature of solidification of the feed stream, and a maximum temperature value T_{max}, corresponding to the temperature of vaporization or the maximum temperature allowed by the membrane:

$$J_b(T) = A_T T^2 + B_T T + C_T; \quad T = (T_{min}; T_{max}) \qquad (2.11)$$

with A_T, B_T, and C_T fitting parameters of this equation.

Although this relationship may enter the model, it is normally not used since the feed stream temperature is controlled, and thus constant, once defined and fixed. Controlling the temperature is an additional cost, and must exhibit some convenience.

At high pressure levels, generally used in NF and reverse osmosis (RO) processes, temperature control may be mandatory in case of batch membrane processes, since the continuous compression and decompression of the feed stream leads to temperature increase. If this is of benefit (that is, J_b increases as a function of T), it may be left uncontrolled up to the maximum allowable temperature of the system, even if Eqn (2.11) is not known. In all other cases, temperature control appears to be mandatory and should be maintained at the boundary conditions known by the membrane process designer.

Membrane Properties

Smaller average pore sizes of the membrane lead to smaller values of J_b due to lower permeability values, but the final value is strongly affected by the feedstock characteristics and membrane surface characteristics: as soon as higher concentrations of suspended particles and/or suspended solids in the dimensional range between 1/10 and 10 times the mean pore size are present, J_b may decrease sensibly. Moreover, boundary fluxes may increase if the membrane surface is phobic to the bulk stream's nature. In other words, hydrophobic membranes will exhibit greater J_b values in contact with aqueous solutions.

There is no need to have some empirical relationship here, since the influence of the specific membrane will be included during the boundary flux measurement experiments and, as a consequence, automatically as fitting parameters in the model.

Time

Whereas critical fluxes remain unaffected and constant, threshold fluxes change as a function of time. Concerning the permeate flux, integrating Eqns (2.5) and (2.6), the following relationships were found:

$$J_p(t) = J_p(0) + \beta^{-1}\left(\alpha + \beta\, J_p(0) - \beta\, J_b\right)\left(\exp\left(-\beta\, TMP\, t\right) - 1\right);$$
$$J_p(0 \div t) > J_b \tag{2.12}$$

Equation (2.12) can be used at boundary flux conditions; in this case it becomes:

$$J_b(t) = J_b(0) - \alpha\, TMP_b(t)t \tag{2.13}$$

where TMP_b is the boundary TMP. In the case of processes exhibiting critical flux behavior, when the value of α is equal to zero, the validity of Eqn (2.13) still holds and the boundary flux value results are independent from time. In the case of systems following threshold flux behavior, the changing values of the boundary flux increase the difficulty of following the boundary curve throughout operation. This makes this parameter difficult to use in engineering tools, which prefer constants as input. On the other side, from Eqn (2.13), it appears that TMP_b remains constant as a function of time. This is in accordance with results obtained by other research groups in past years [24,27−29]. Whereas the critical flux or the critical TMP are taken into consideration for process design purposes, since constancy of the linear relationship between these two parameters is always guaranteed, this is not the case as soon as boundary fluxes (and therefore including threshold fluxes as a possible case study) are used. It appears to be preferable to use the value of TMP_b as an input parameter for the design tools.

Feedstock Characteristics

Concerning the feedstock characteristics, this parameter is sensibly affected by pretreatment processes. A good design of the pretreatment processes (so-called pretreatment tailoring) before the membrane section appears to sensibly increase the value of J_b. Correct pretreatment tailoring for membranes is a difficult task, since the modification that occurs on the feed stream must be valuable for all successive membrane sections. Some researchers use microfiltration ("MF") as pretreatment for the subsequent membrane steps [30,34]. Although this approach saves ultrafiltration ("UF") and nanofiltration ("NF") from sensible fouling, MF sustains a heavy duty. Fouling is totally formed and shifted by one separation step, and make the overall process difficult. Using ceramic or tubular modules makes the cleaning of membranes easier, but still does not solve the problem due to high cleaning procedure costs [34,35]. The best strategy is therefore to distribute fouling among all involved membrane steps equally. Pretreatment processes should therefore be of a physical or chemical nature. As an example, the application of fungi for organic matter reduction works well for UF, but due to the production of enzymes of a size near the NF pore size, on this latter membrane the threshold flux values sensibly drop [29]. As a consequence, this pretreatment does not fully qualify for the process. Better results may be obtained by adopting microalgae, but this process is still under investigation [16]. Moreover, membrane bioreactors have been investigated, with promising results, but they are characterized by insufficient productivity [36,37]. The same problem affects membrane distillation-based processes [38]. Flocculation by heavy metal salts efficiently increases threshold flux values on all membranes, but in case of olive mill wastewater (OMW) treatment, heavy metal ions still remain and are measured in the reverse-osmosis permeate: again, the pretreatment does not qualify since it fails to undercome the legislative limits for wastewater recovery [31]. This is not a general rule: in cases of treating by flocculation tomato vegetation wastewater, tannery wastewater, or marine sediments, these problems were not encountered [25,26,39]. Flocculation was therefore developed by substituting the flocculent with nitric acid. The obtained benefit in threshold flux value increases was reduced, and the problem of heavy metal ions in the permeate stream was overcome [18]. Photocatalysis appears to be less efficient than flocculation, but it increases the threshold flux values of all membranes and allows production of a purified water stream compatible to the sewer system due to organic matter reduction and organic chain cutting. Photocatalysis was studied both as a Fenton process and by using a titania nanocatalyst [40−45]. The problems here are the high operating costs of the reactants, making the process economically unfeasible. The solution here was to use a magnetic core nanocatalyst, which can be completely recovered by a magnetic trap and reused in successive batches [46−50].

The optimized application of different pretreatment processes inhibits differently the triggering of the fouling processes over the membrane surface, and fouling indexes may enter a technically sustainable range [45]. By using OMW, only threshold flux and no critical flux values were found and measured. The data can be used for process optimization purposes [18]. This permits maintenance of membrane efficiency for a very long period of time [20].

In all these works it was observed that fouling is mainly given by both dissolved and suspended solutes, and since tracking these parameters is normally a difficult task, a key parameter KP must be chosen to define a fingerprint of the feedstock, useful to fit to J_b values.

The authors have proposed a fitting curve of J_b, based on the general relationship between the permeate flux J_p and TMP, the membrane permeability, and the osmotic pressure as a function of KP and the time t, that is [51]:

$$J_b(KP, t) = w\, P_b(0) - \alpha\, t\, P_b(0) - \left[w\, p_1 - \alpha\, p_1 t + m_1 P_b(0)\right] KP + m_1 p_1 KP^2$$

$$(2.14)$$

where $P_b(0)$ is the applied operating pressure at the boundary flux conditions at the start of operation, in detail:

$$P_b(0) = TMP_b + \pi(KP)\big|_{t=0} \qquad (2.15)$$

Again, from Eqn (2.15), TMP_b appears to be constant as a function of time.

In Eqn (2.14) w, m, and p are fitting parameters of the respective equations fitting the membrane permeability m and the osmotic pressure π to KP, respectively:

$$m(KP, t) = w - m'\, KP \qquad (2.16)$$

$$\pi(KP) = p\, KP \qquad (2.17)$$

The pure water permeability, that is w, may be a function of time depending on the amount of irreversible fouling formed on the membrane.

The obtained relationship (Eqn (2.14)) is a second-order polynomial equation, valid in the physical range of $KP = [0, +\infty)$, and qualifies as a better fitting equation, since:

1. The limit now has a physical meaningful limit at pure water conditions, that is $\lim [KP \to 0]\, J_b(KP, t) = w(t)$.
2. The fitting curve is always convex, since $(m'\, p) > 0$, and always has two roots. The first one will hereafter be called KP*, which represents the upper limit of solute concentration in the feed solution to trigger almost instantaneously zero flux conditions.

3. For [t → ∞], the minimum point tends to − ∞; in other words, the relevant $J_b(KP)$ values as a function of time become lower and lower, and the membrane is less productive. There is a time point t* for a given KP where zero flux conditions are immediately met, and thus the module is completely dead. In fact, the reason for this is the irreversible fouling buildup and/or aging of the membrane.

4. It appears that the operating boundary pressure P_b (and as a consequence, if KP is fixed, the value of TMP_b) is not a function of time. This particular behavior was previously observed in some works by the authors [19,29].

5. Although at high KP values the fitting curve tends to positive J_b values, this part of the curve has no physical meaning; therefore, the validity of the fitting curve must be restricted to KP values in the range [0,KP*].

Figure 2.2 summarizes all the characteristic points of the new fitting equation, that is, Eqn (2.14).

Again, the boundary flux includes all critical and threshold flux concepts under one common name, as schematically shown in Figure 2.3. In the same Figure, the plot of typical flux profiles corresponding to the relevant boundary flux types are shown. It should be noticed that the value of m affects the initial profile of the flux curve, in the very first moments, and thus affecting the maximum possible productivity of the membrane, whereas α and β affects the shape of the flux profile after a longer period of time determining a flux reduction. Since scope of this Handbook is to maximize longevity of the membrane modules, the latter region is the one of interest here.

It may be noticed that the critical flux appears to be a subclass of the threshold flux, and that as a consequence the introduction of the boundary flux appears redundant. On the other hand, it is the opinion of the authors that the distinction between critical and threshold flux should be maintained, for two reasons: the first is to keep intact credit for the critical and threshold flux concepts to those researchers who first defined it; and the second is that there is a main difference in the behavior of the system if it follows a critical or threshold flux relationship, that is, time. Critical fluxes have a great advantage of not changing as a function of time once measured; in other words, once measured, the same critical flux may be adopted when considering different

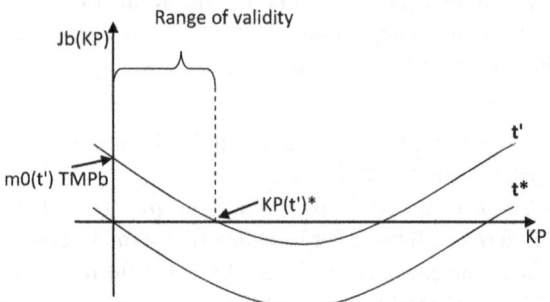

FIGURE 2.2 Plot of Eqn (2.14) at a fixed time point t′ and t*.

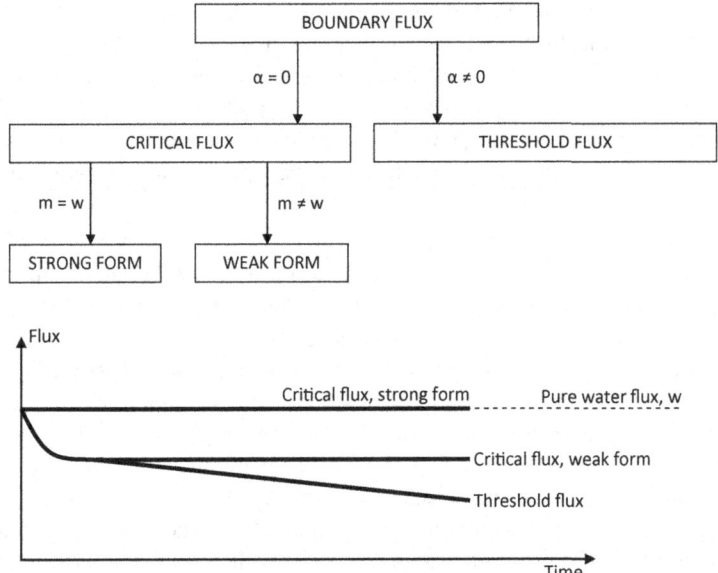

FIGURE 2.3 Relationship between the critical, the threshold, and the boundary flux.

operating times. The same is unfortunately not true for threshold fluxes, since values change as a function of time, and thus require determination by both measurements and correlation functions. For this reason, the distinction between critical and threshold flux appears important and should be maintained, although it may be merged in a common term (that is, boundary flux) for a more practical use of the concepts.

Cleaning of Membranes

The cleaning of the membranes may be performed by different methods:

1. Rinsing
2. Washing
3. Osmotic cleaning

and procedures:

1. On the feed side
2. By backwashing
3. Air bubble-assisted
4. Other methods

In all cases, the target is to eliminate the semi-reversible fouling of the membrane. The definition is strictly related to the adopted cleaning procedure, since the fouling that is not eliminated by the cleaning medium must be accounted as irreversible fouling.

In all cases, the cleaning of the membrane leads to an increased or equal value of the permeability and the pure water permeability. The best practice is to check the efficiency of the cleaning procedure by pure water permeability values. If w_1 and w_2 are the pure water permeability values after two subsequent cleaning procedures separated by a period of time equal to τ, measured before normal operation, the cleaning efficiency $\Delta w\%$ can be calculated as:

$$\Delta w\% = \left(w_1 - w_2\right)\left(\tau\, w_1\right)^{-1} \qquad (2.18)$$

If no irreversible fouling was formed, working at sub-boundary flux conditions, the value of $\Delta w\%$ is observed to be constant.

Cleaning can be always performed on the feed stream side, and normally this is performed by allowing zero or a small amount of permeate to pass through the membrane. If the module allows backwashing, such as hollow fiber modules, pressure may be applied to the permeate side of the membrane, allowing higher possible inverse permeate flow rates to cross the membrane. Backwashing is one of the most efficient procedures, since it effectively cleans the membrane pores.

The rinsing of the membrane is normally performed by the permeate stream or water. The adopted cleaning medium is cheap but may not guarantee high $\Delta w\%$ values.

In cases where the addition of chemicals is performed, the cleaning procedure is referred to as washing of the membranes. Chemicals increase the efficiency of the remediation procedure, but they have additional costs. In many cases, acqueous solutions with the addition of some acid or base is used. This common practice represent the standard procedure for membrane washing, and as a consequence, the determination of the part of fouling being semi-reversible.

The two methods can be performed by different procedures, in most cases by recycling the cleaning medium over the membrane for a certain period of time. One or more cleaning solutions may be used during membrane washing. The cleaning of the membrane may be repeated if the measured cleaning efficiency value ε is not satisfying or as expected.

A third method is osmotic cleaning. In this case, the process is simply stopped for a certain period of time. The permeate, more diluted than the feed stream, will diffuse back to the feed side. The advantages of this method are high cleaning efficiencies, use of no chemicals, and the fact that the method can be adopted as a backwash method on membranes that do not support this cleaning procedure, such as spiral-wound ones. The main disadvantage of this method is that it is extremely costly in terms of time, far away from common industrial practice and needs.

Rinsing and washing may be assisted by air bubbles. In this case, the presence of changing phases in the feed stream gives rise to additional turbulent effects, which aid in the detachment of semi-reversible fouling layers. The method is well established in membrane bioreactor (MBR), where biomass flocks on the membrane surface are efficiently detached by air bubbling.

Chapter 3

The Boundary Flux Model

The simulation software should integrate Eqns (2.5)–(2.9) and (2.15) as a function of time, and uses the following relationship to estimate the selectivity on the key parameter KP:

$$R(KP) = \sigma\, TMP(TMP + \gamma)^{-1} \qquad (3.1)$$

where σ, called the reflection coefficient, and γ are two fitting parameters.

At the University of Rome "La Sapienza," Department of Chemical Engineering, the developed model was called "Memphys" ("*MEMbrane Processes HYbrid Simulator*"); it requires input parameters to work correctly, as reported in Table 3.1.

MEASUREMENT OF THE MODEL INPUT PARAMETERS

The model requires some data as input to work correctly. The determination of this data may require proper experimental work, which is suggested to be carried out before plant design.

Pure Water Permeability and Cleaning Efficiency

The measurement of the pure water permeability of a membrane is easy to carry out. The membrane module is fitted in the pilot plant, and deionized water is used as feedstock. The experiments are performed immediately after a membrane cleaning procedure, in batch mode and returning the permeate stream back to the feedstock. The permeate line should be as short as possible to approximate the pressure on the permeate side equal to atmospheric. Temperature must be controlled and held constant. The pure water permeabilities on both new and used membrane modules can be measured.

Three points at different pressure values (transmembrane pressure, TMP) may be sufficient. The respective permeate flow rate J_p is noted, checking for flow rate stabilization for at least 5 min.

The obtained data points are fitted by a fitting curve; generally the following equation should hold:

$$w = J_p TMP^{-1} \qquad (3.2)$$

This measurement may be integrated as a periodic check in an advanced control system.

The Boundary Flux Handbook. http://dx.doi.org/10.1016/B978-0-12-801589-6.00003-6

TABLE 3.1 Input Parameters Required by the Simulation Software

Input Parameter	Description	Determination
A	Membrane area	User defined, must meet the project requirements in terms of F^*_p
σ	Fitting parameters concerning selectivity	KP versus TMP plot
γ		
p	Fitting parameter of the osmotic pressure	J_p versus TMP plots at different KP values
m'	Fitting parameter of membrane permeability	J_p versus TMP plots at different KP values
J_b	Boundary flux value	Boundary flux measurements
TMP_b	Operating boundary pressure	
α	Sub-boundary fouling rate index	
w	Pure water permeability	Pure water permeability measurement
KP(0)	Starting value of KP	Chemical or other analysis
V(0)	Starting feedstock volume	User defined
$\Delta w\%$	Cleaning efficiency	Pure water permeability measurement at the start of operation between subsequent cleaning cycles

The cleaning efficiency value is evaluated normally after many subsequent cleaning procedures by observing different pure water permeability values, by adopting Eqns (2.19) and (3.2).

Permeability and Osmotic Pressure

The determination of the permeability values requires extended experimental work. The membrane module is fitted in the pilot plant, and some feedstock at different and known KP values must be available. The feedstock should be able to cover the expected range of KP values during normal operation. At least three different feedstocks, characterized by a minimum, a maximum, and a midpoint value of KP, must be used.

The experiments are carried out in batch mode and the permeate is returned back to the feedstock. The permeate line should be as short as possible to approximate the pressure on the permeate side equal to atmospheric. The temperature must be controlled and held constant. The permeabilities on both new and used membrane modules can be measured.

Three points at different absolute pressure values P for each feedstock may be sufficient. The respective permeate flow rate J_p is noted, checking for flow rate stabilization for at least 15 min.

The data can be used for the determination of both the osmotic pressure π and the permeability m, by a linear fitting:

$$J_p(P, KP) = m(KP)[P - \pi(KP)] \tag{3.3}$$

Once the different values of $\pi(KP)$ and $m(KP)$ are obtained, the values can be fitted by Eqns (2.17) and (2.18).

Selectivity

The selectivity on the chosen key parameter KP should be expressed as a function of TMP by Eqn (3.1). The experiments are not so easy to carry out and require time. A representative feedstock of the system is required.

The experiments are carried out in batch mode and the permeate is returned back to the feedstock. The permeate line should be as short as possible to approximate the pressure on the permeate side equal to atmospheric. The temperature must be controlled and held constant. Selectivities towards KP on both the new and used membrane modules can be measured.

As many TMP values as possible should be investigated, concentrating data points near the osmotic pressure. At least one measurement should be performed to the maximum allowable TMP value of the system, as well.

Once the analyses are performed on both the feed and permeate streams, rejection is calculated as:

$$R(TMP) = 1 - (KP_{PERMEATE} KP^{-1}_{FEED}) \tag{3.4}$$

Boundary Flux

Equation (2.5) can be integrated between a time point t_1 and t_2, and the following linear equation can be derived:

$$m(t_2) - m(t_1) = \Delta m = \alpha(t_2 - t_1) \tag{3.5}$$

Permeate flux and permeability values are strictly connected by the following general equation:

$$m(t) = J_p(t)/TMP(t) \tag{3.6}$$

Merging together Eqns (3.5) and (3.6), the following relationship is obtained:

$$J_p(TMP, t_1) - J_p(TMP, t_2) = -\Delta J_p{}^* = \alpha\, TMP(t_1 - t_2) \qquad (3.7)$$

which is valid in cases where the same TMP value is used at t_1 and t_2. It is possible to use different TMP values between t_1 and t_2 without invalidating Eqn (3.7): as long as the adopted TMP values remain below the boundary one, no effect on changes of the permeability loss rate should be observed. $-\Delta J_p{}^*$ is the expected permeate reduction if Eqn (2.5) holds (that is at subthreshold flux regimes) and must be compared to the measured one equal to $-\Delta J_p$.

The boundary flux can be measured by different procedures, as follows:

1. Flux-pressure profiles
2. Flux or pressure cycling
3. Flux or pressure stepping

All methods rely on starting the measurement in sub-boundary operating conditions, therefore near to zero flux values or the osmotic pressure. Moreover, polarization must be fully developed: the membrane should work for hours at low flux in contact with diluted feedstock before the experimental campaign even starts.

The first method is very simple, but not as rigorous as the other two methods. It consists of checking at which flux value the linearity to the pressure is lost. With increases in pressure—and, as a consequence, the permeate flow rate—in subthreshold conditions, the gain should remain proportional. As soon as the boundary flux value is exceeded, this behavior changes.

The pressure cycling method was initially proposed by Espinasse et al. [52] for critical fluxes. Basically, the method consists of cycling the applied pressure up and down by a constant TMP variation and equal to ΔTMP, and to check for the reproducibility of the permeate flux at same pressure values before and after the pressure changes (see Figure 3.1).

The lowest pressure value at which the difference between the measured $-\Delta J_p$ and the evaluated value from Eqn (3.7) $-\Delta J_p{}^*$ in the same period of time becomes positive is the boundary pressure TMP_b, and the boundary flux value is determined by taking into account the permeate flux value at the beginning of the correspondent pressure cycle (see Figure 3.2).

In order to use this method for boundary flux measurements, the value of the sub-boundary fouling index α must be determined. The determination of the value is performed during the first cycle, which has the greatest probability to be sub-boundary, by the following relationship:

$$\alpha = -\Delta J_p{}^*[TMP(t_1 - t_2)]^{-1} \qquad (3.8)$$

The next pressure cycle is required to confirm that the first one was effectively performed in sub-boundary conditions. If this is not the case, a lower value of ΔTMP must be chosen.

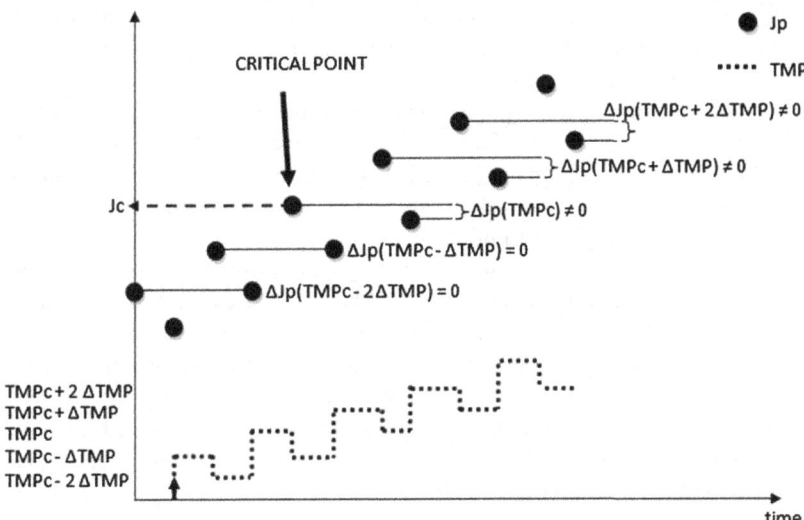

FIGURE 3.1 Critical flux determination by application of the pressure cycling method.

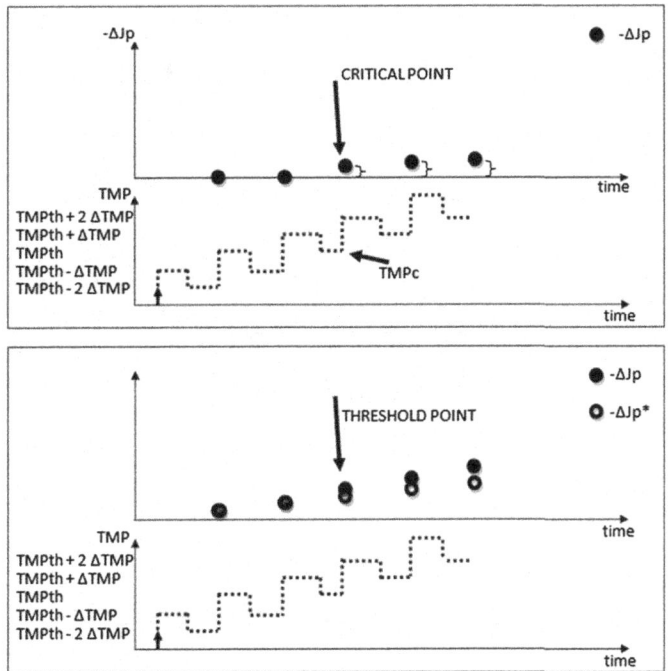

FIGURE 3.2 Critical (top) and threshold (bottom) flux determination by a $-\Delta J_p/-\Delta J_p^*/$ TMP—time plot.

The flux cycling method is similar, targeting specific flux values in series by adopting different pressure values. The basic strategy is the same as used for the pressure cycling method.

As soon as the boundary point is identified, the best practice is to confirm the observation by one more flux or pressure cycle and then to immediately stop the experimental campaign, in order to avoid working for long periods of time in super-boundary flux conditions and, as a consequence, forming irreversible fouling over the membrane.

The third method, that is, the pressure stepping method, is similar to the cycling one, although the increases of ΔTMP are subsequent up to a value of maximum investigated TMP value, and after this ΔTMP decreases are applied. The main disadvantage of this method is that measurements can be performed for a long period of time in super-boundary flux conditions and, as a consequence, lead to sensible fouling over the membrane. Moreover, if this is the case, the measured boundary flux value may be different from the effective one after the fouling-promoting experimental campaign.

Chapter 4

Process Control

MEMBRANE PROCESS CONTROL STRATEGIES

Two main membrane process control strategies can be adopted: controlling the permeate flow rate by changing the applied pressure value via a regulation valve on the concentrate line, or directly controlling the constancy of the applied pressure. All strategies normally include temperature and flow rate control of the feed stream; thus, generally Eqns (2.10) and (2.11) are not required.

In the next figures, the temperature control system is not represented but should require a temperature sensor and heat exchangers to work. The feed stream control system depends on the nature of the pump used. In the case of centrifuge pumps, speed control or regulation valves at the pump inlet may be sufficient. In the case of volumetric pumps, a by-pass or a back-recycle line on the pump should be provided.

In any case, the control of the two parameters requires attention to how to perform the control; and in some cases advanced control systems such as decouplers must be used. Details may be found elsewhere and are out of the scope of this handbook.

Control Strategy 1: Constant Permeate Flow Rate

The advantage of this control strategy is the possibility of always maintaining the productivity of the membrane plant in line with the project values, referred to hereafter as F^*_p. Moreover, the application requires the use of a simple proportional integral (PI)-type control system.

The challenge of this strategy is to correctly define the permeate flow rate set-point of the controlled system. Since boundary flux values will change as a function of time and of the pollutant concentration in the feedstock, it is not correct to assume as set-point value the boundary flux rate value at the beginning of the operation. In particular, this problem would have higher impact when applied to batch membrane processes, as reported in Figure 4.1.

The Boundary Flux Handbook. http://dx.doi.org/10.1016/B978-0-12-801589-6.00004-8

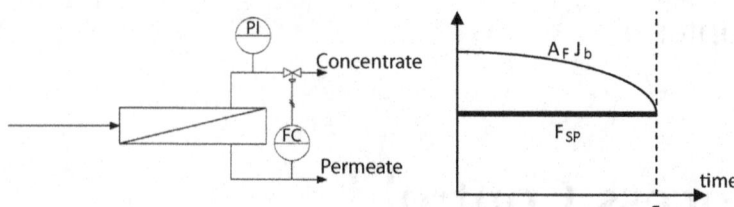

FIGURE 4.1 Control scheme of Strategy 1 and relevant J_b versus t plot.

Control Strategy 2: Constant Operating Pressure Values

The advantage of this control strategy is the simplicity, but the productivity of the membrane plant may not be in line with the expected project values, in particular permeate flow rate is not constant.

As noted earlier in this handbook, the boundary TMP_b value is not a function of time, and thus it is constant during operation. The challenge of this strategy is to define correctly the operating pressure set-point P^* of the controlled system: the osmotic pressure will change as a function of time and of the pollutant concentration in the feedstock, and needs to be added to the TMP value. In particular, this problem would have higher impact when applied to batch membrane processes, as reported in Figure 4.2. Nevertheless, the set-point can be defined by analysis of the starting feedstock conditions, and as a consequence, calculations can be performed with ease when compared to the previous control strategy.

Comparison of the Two Proposed Control Strategies

Some observations on applying the control of the permeate flow rate (FC) or of the operating pressure (PC) to both continuous and batch membrane processes are listed herein, partly in accordance with previous findings from Vyas et al. [53].

- The control of the process is performed in both cases by a PI-type controller.

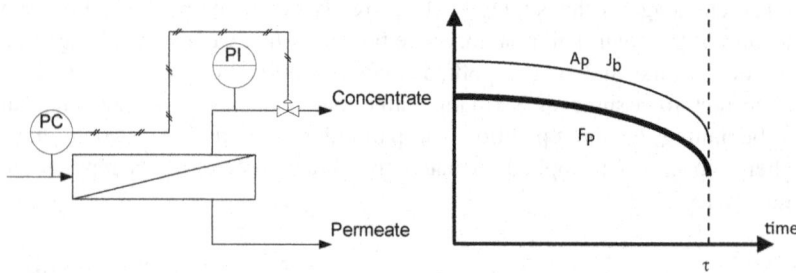

FIGURE 4.2 Control scheme of Strategy 2 and relevant J_b versus t plot.

- The control of the process by the FC strategy guarantees constant productivity of the plant, in particular for continuous processes. This factor is important as soon as subsequent downstream processes require constant permeate flow to work properly. The PC strategy cannot guarantee this result of constancy, and requires the presence of some accumulation tank or the membrane process on the permeate side should not have any other down stream processes.
- The membrane area requirement is equal or lower in case of adopting the PC strategy if compared to the FC strategy. This has a direct influence on the investment costs, too.

Therefore, the best control strategy depends on the membrane system and the feedstock characteristics. In cases in which the feedstock at the start of operation is always constant and the osmotic pressure changes may be neglected (that is, for values of p_1 equal to or near zero), the PC strategy may be advantageous concerning ease of control and membrane area requirements. In all other cases, the FC strategy is suggested.

Chapter 5

Design and Optimization Guidelines of Continuous Membrane Processes

The optimal operation and control strategy for membrane processes will be extended to membrane fouling: it appears mandatory to operate at or below the boundary flux in order to strongly inhibit irreversible fouling issues.

The design of a membrane process will follow this guideline. Since calculations are complex and should be performed by proper simulation software, the simplified approach presented here will always tend to determine the parameters on the safe side. As a consequence, the draft design produced by this approach is useful for a first tentative design of the process and to evaluate the technical and economic feasibility, but such design merits further optimization by simulation software in a successive second step. This is not a problem, since this handbook aims to provide tools for initial counseling on membrane technologies, and thus the estimation design of both technical and economic aspects of the process in worst-case conditions fits well with this aim. Please note that all of the following calculations must be performed by using the units reported in the variables list chapter.

The initial project inputs are the minimum desired rejection factor R^* and minimum total productivity F^*_p in terms of total permeate flow rates.

The first check should be performed on selectivity at an expected transmembrane pressure (TMP) value. For each membrane, depending on the feedstock and the expected key parameter (KP) values, boundary flux J_b and pressure values TMP_b must be known either from proper experimental campaigns on lab and pilot plants or from a bibliographic source, such as the included database in this handbook. As soon as $R^* < R(KP,TMP_b)$, the chosen membrane qualifies for optimization procedures. If more than one membrane meets this condition, further evaluation must be performed in order to make a final choice. As a general rule, in this case the membrane exhibiting the highest permeability should be initially chosen.

In addition to this, at boundary conditions, the feed stream velocity over the membrane v_F and the operating temperature T should also be provided and kept constant. A second check is performed on the sustainability

The Boundary Flux Handbook. http://dx.doi.org/10.1016/B978-0-12-801589-6.00005-X

concerning the recovery of permeate. A limit in adopting continous membrane processes is the maximum allowable recovery factor value: if set too high, the consequence is to have insufficient velocities of the concentrated stream exiting the membrane modules capable to inhibit fouling. Indeed, a main constraint for the use of continuous membrane processes is the continuous availability of the feed stream, which should be kept almost constant in terms of flow rate and quality. This result can be achieved either by constant performing up stream processes or the availability of a huge feedstock, such as seawater in case of desalination. If this is not the case, batch processes applies best. Therefore, this check must be performed on the velocity of the concentrate exiting the last module, since it should be at least equal to the ones the boundary flux is known. If A_P is the feed stream passage area of the membrane module, F_F the maximum feed flow rate of the adopted pump, and v_F the minimum desired concentrate stream velocity over the membrane surface at the outlet and equal to that of the chosen boundary conditions, then N_M, that is, the parallel membrane module lines connected to the pump, is equal to:

$$N_M = 0.001 \left(F_F - F^*_p\right) A_p^{-1} v_F^{-1} \qquad (5.1)$$

The result should be rounded to the lower unity. The value of N_M must always be equal or greater than 1. In any case, the value of F_F should be much higher than the value of F^*_p. This condition represent a necessary hypothesis to further calculations here presented in this handbook. If this is not the case, a pump with a higher capacity must be chosen. In these conditions, the recovery factor Y^* is evaluated as:

$$Y^* = F^*_p F_F^{-1} \qquad (5.2)$$

Depending on the system and general process requirements, finally the control strategy must be chosen.

In the case of adopting an FC strategy (control of the permeate flow rate), simulation software is required to estimate the boundary flux value at the exit of the membrane series. The estimated boundary flux value at the exit of the last membrane module, depending on the target recovery factor Y, which may be lower than or equal to Y*, reduced by a safety margin δ_F of 5–10%, is finally equal to the project permeate flux value, which serves as set-point to the permeate flux controller.

In these conditions, the concentrate will have a characteristic value of KP_c equal to:

$$KP_c = KP(0)(1 - Y^* + Y^* R(TMP_b))(1 - Y^*)^{-1} \qquad (5.3)$$

The required membrane area A' can be evaluated by the following relationship:

$$A' = F^*_p (1 - \delta_F)^{-1} J_b (KP_C)^{-1} \qquad (5.4)$$

In order to consider the loss of permeability after every membrane cleaning procedure, if C is the number of desired separation cycles lasting τ hours, the final required membrane area A_F becomes:

$$A_F = A'(1 + (C - 1)\tau \, \Delta w\%) \tag{5.5}$$

The set-point of the feed flow rate should be set to:

$$F_{SP} = F^*_p \tag{5.6}$$

By this approach, once a set-point value of the permeate feed flow rate controller is defined and after the correct estimation of the required membrane area to use to reach the target performances, a simple proportional integral (PI) control system can be used. Moreover, this approach guarantees operation of the membrane plant far away from the boundary condition for most of the time. Adjustments to the set-point value are needed only in case of a changing feedstock: in this case, a fingerprint in terms of KP of the feedstock should be monitored as input to an inferential advanced control system, capable of evaluating proper permeate flow rate set-point values by Eqn (2.15).

If the adopted strategy is of the PC type (control of the operating pressure), the first step is to consider the boundary TMP_b value, reduced by a safety margin δ_P of $5-10\%$. The required membrane area A_P can be evaluated by the following relationship:

$$A_P = F^*_p (J_b(0) - 0.5 \, \alpha \, TMP_b (1 - \delta_P)\tau)^{-1} (1 + (C - 1)\tau \, \Delta w\%) \tag{5.7}$$

In general, the set-point operating pressure value is equal to:

$$P_{SP} = TMP_b (1 - \delta_P) + \pi(KP)|_{t=0} \tag{5.8}$$

By this approach, once a set-point value of the operating pressure is defined and after the correct estimation of the required membrane area to use to reach the target performances, a simple PI control system can be used. Adjustments to the set-point value are required by changes of the feed stream: in this case, a fingerprint in terms of KP of the feedstock should be monitored as input to an inferential advanced control system, capable of evaluating the osmotic pressure and the correct P_{SP} values by Eqn (2.15) at the start of operation.

Summarizing, the advantages and disadvantages of the FC and PC strategy are briefly described as follows:

- The membrane area requirements in the case of continous processes are equal or lower for PC strategies, if compared to FC strategies.
- The control of the process is performed in both cases by a PI-type controller. In the case of changing feedstock at the start of operation, advanced control is necessary: PC strategies require the indirect measurement of the initial osmotic pressure of the feedstock, and this is in most cases not an easy task to accomplish; FC strategies require the simulation software to calculate a new value for F_{SP}. In both cases the purpose is to give a different set-point to the PI-type control systems.

- The simulation software is used in both cases to estimate the required membrane area. Only in the case of FC strategies the simulation software need integration within the control system, in case of changing feedstock at the start of operation. This is not the case when adopting a PC strategy, since set-point calculations can be performed on the basis of the conditions at the start of operation.

The number of membrane modules is defined by the area of each module, equal to A_m, by the following relationship:

$$N = AA_m{}^{-1} \qquad (5.9)$$

where A is equal to A_F or A_P in the case of adopting an FC or PC strategy, respectively. The obtained number of membrane modules should always be rounded to the upper unity. By adoption of this procedure, selectivity and productivity are on target, while maximum longevity is guaranteed, but at this point the process is designed for all membrane modules working at the same operating conditions, and this requirement is met as long as no membrane is put in series—that is, as long as $N \leq N_M$. It is up to the discretion of the designer whether to accept this value or to try to put membrane modules in series by lowering F_F, that is, by using a pump characterized by a lower flow rate capacity.

As soon as $N > N_M$, the membrane process module layout must be studied. With N_M the number of the membrane module parallel lines connected to one pump and N_S the membrane modules in series on each line, the following equation should, at the end, hold:

$$N_S = NN_M{}^{-1} \qquad (5.10)$$

The value of N_S, again rounded to the upper unity, depends on the pressure drop across a single membrane module of the series ΔP_M. If a precise value is not known from the supplier, some typical values reported in Table 5.1 can be applied.

Because starting from the entrance of the first membrane module boundary flux conditions must be attained, the TMP values at the exit and in every

TABLE 5.1 Typical Values for ΔP_M (Feed Flow Rate Equal to $2000 \, l \, h^{-1}$)

Membrane Module	ΔP_M (bar m^{-1})
Plate and frame	0.1
Spiral wound	0.6 ÷ 1.2
Hollow fiber	0.3 ÷ 0.7

successive membrane module are lower, thus leading to lower rejection values and permeabilities.

Three conditions need to be checked.

The first check is performed on the pressure. The following equation must be valid:

$$TMP_b - N_S \Delta P_M < 0 \tag{5.11}$$

The second check is on the rejection, calculated at the end of the membrane series, and the following condition must hold:

$$R^* < R(TMP_b - (N_S - 1)\Delta P_M) \tag{5.12}$$

The third check is on permeabilities. Again, the permeability m is a function of KP and, in the last instance, of R(KP). The procedure is to calculate in loop the following equations, seen beforehand on one membrane module, for the series, from the first module of the series ($i = 1$) to the last one ($i = N_S$):

1. $m_i(KP(i - 1))$ by Eqn (2.17)
2. $TMP_i = TMP_b - (i - 1)\ \Delta P_M$
3. $J_{p,i} = m_i\ TMP_i$
4. $F_{p,i} = J_{p,i}\ A_m$
5. $F_{F,i} = F_{F,i-1} - F_{p,i}$
6. $Y_i = F_{p,i}\ F_{F,i}^{-1}$
7. $KP(i) = KP(i - 1)\ (1 - Y_i + Y_i\ R(TMP_i))\ (1 - Y_i)^{-1}$
8. $F_{p,T} = \Sigma_i\ F_{p,i}$

where $F_{F,0} = F_F\ N_M^{-1}$

All variables used during this procedure uses the same defined beforehand, with the addition of an index i to refer to the correspondent step. At the end of this procedure, the condition to check is:

$$F^*_p < A'A^{-1}N_M F_{p,T} \tag{5.13}$$

If all conditions are met, the draft technical design is completed successfully.

If any of the conditions are not met, the entire procedure must be checked again by increasing N_S by one unit or starting from Eqn (5.1) using higher-flow-rate pumps with increased F_F values or splitting F^*_p among separated membrane systems served independently by N_P pumps. In the worst case, the considered membrane must be changed, which requires the determination of all involved parameters again.

The optimization of economics must be evaluated by calculating the best values of N, τ, and C in order to minimize the main costs of the process $.

The guidelines presented here represent a simplified method to perform a basic economics check used to compare different membrane processes and layouts, since some expenses are not considered, such as personal costs and extraordinary maintenance.

For this procedure, concerning the membranes, the unity membrane module cost $m and the unity housing cost $h hosting M modules must be known. Moreover, the operating life Φ of the plant must be fixed. The units of Φ must be the same as of τ.

In first instance, concerning membranes only, the initial capital costs are equal to:

$$\$c^* = N \,\$m + NM^{-1} \$h \qquad (5.14)$$

Once the membrane modules are depleted and replacement is needed, all membrane modules must be substituted. During the operating lifetime of the plant, the following hardware costs must be considered:

$$\$c' = \$c^* \left(1 + \Phi \, \tau^{-1} C^{-1} \right) \qquad (5.15)$$

If N_P is the number of adopted pumps, cp the unity cost, and τ_P the expected lifetime, the relative investment costs can be evaluated as:

$$\$c'' = N_p \$cp \, \Phi \, \tau_p^{-1} \qquad (5.16)$$

If the pump efficiency η is fixed, normally assumed to be around 0.7, the operating costs as a function of the electricity costs $e for each kWh may be estimated as equal to:

$$\$o = 2.78 \times 10^{-6} \, N_P \, F_F \, \rho \, g \, TMP_b \, \Phi \, \$e \, \eta^{-1} \qquad (5.17)$$

where ρ is the feed stream density expressed in kg L^{-1} and g is equal to 9.81 m s^{-2}.

The main total costs $ are therefore:

$$\$ = \$' + \$'' + \$o \qquad (5.18)$$

Eqn. (5.18) requires minimization as a function of the parameters N, τ, and C by additional software, such as Excel or Memphys [38]. If the total costs are too high, most probably the project target values, such as Y*, were set too high and require revision.

Chapter 6

Design and Optimization Guidelines of Batch Membrane Processes

The guidelines for batch membrane processes follow those presented in the previous chapter on continuous operation. The additional difficulty lies in the changing feedstock characteristics due to batch concentration phenomena and, as a consequence, in the changing values of J_b. The main advantage of using batch processes is the possibility to target higher values of Y*, in case of limited feedstock volumes, since the constraint on the available feed flow rate is neglected. Please note that all of the following calculations must be performed by using the units reported in the variables list chapter.

The initial project inputs are the minimum desired rejection factor R*, the recovery factor Y*, and the desired operation time for one batch θ*.

The first check should be performed on selectivity at an expected transmembrane pressure (TMP) value. For each membrane, depending on the feedstock and the expected key parameter (KP) values, boundary flux J_b and pressure values TMP_b must be known either from proper experimental campaigns on lab and pilot plants or from a bibliographic source, such as the included database in this handbook. As soon as R* < R(KP,TMP_b), the chosen membrane qualifies for optimization procedures. If more than one membrane meets this condition, further evaluation must be performed in order to make a final choice. As a general rule, in this case the membrane exhibiting the highest permeability should be initially chosen.

In case of FC strategy, once the value of the desired recovery Y* has been fixed, by means of a simulation code, such as Memphys, the following operating strategy is suggested: the boundary conditions must be reached just at the end of the operation, as shown in Figure 6.1, at a constant flux rate J_p^* [12].

If a simulation code is not available, the estimation of the final value of KP, starting from a value of KP(0) in the feedstock, should follow Eqn (5.3) in worst-case conditions (that is, in total rejection conditions of KP):

$$KP_c' = KP(0)(1 - Y^*)^{-1} \tag{6.1}$$

The Boundary Flux Handbook. http://dx.doi.org/10.1016/B978-0-12-801589-6.00006-1

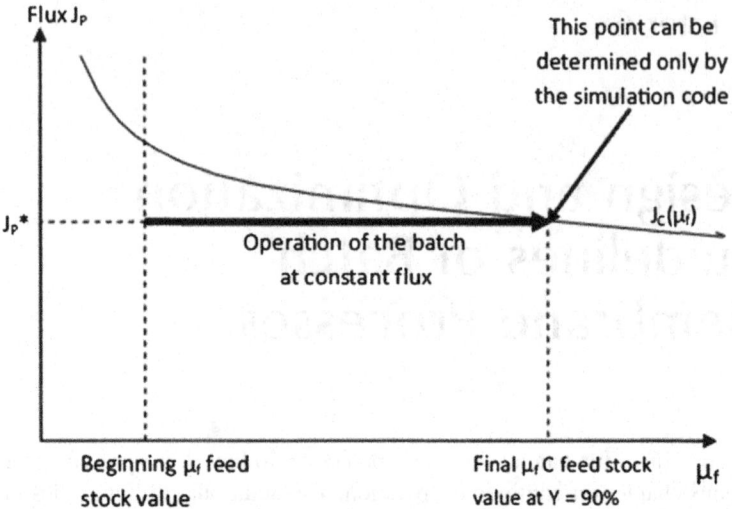

FIGURE 6.1   Suggested operating strategy (in this example $Y^* = 90\%$, KP = electroconductivity μ).

F_p^* is defined by:

$$F_p^* = VY^*\theta^{*-1} \tag{6.2}$$

where V is the feedstock volume and θ^* is the desired operation time for one batch, generally equal to τ. This is not a problem, since this handbook aims to provide tools for initial counseling on membrane technologies, and thus the estimation design of both technical and economic aspects of the process worst-case conditions fits well with this aim.

A condition must be checked on the compatibility of the feasability of the desired operting time θ^*, that is:

$$J_b(0) - \alpha\, TMP_b\, \theta^* > 0 \tag{6.3}$$

If this equation is not valid, the value of θ^* was set too high and must be reduced.

In addition to the above, at boundary conditions, the feed stream velocity over the membrane v_F and the operating temperature T should also be provided and kept constant. A second check is performed on the sustainability concerning the recovery of permeate. A limit in adopting batch membrane processes is the maximum allowable recovery factor value: if set too high, the consequence is to have insufficient velocities of the concentrated stream exiting the membrane modules capable to inhibit fouling. This check must be performed on the velocity of the concentrate exiting the last module, since it should be at least equal to the ones where the boundary flux is known. If A_p is the feed stream passage area of the membrane module, F_F the maximum feed flow rate of the adopted pump, and v_F the minimum desired concentrate stream velocity over the membrane surface at the outlet and equal to that of the chosen boundary conditions, then N_M, that is, the parallel membrane module lines connected to the pump, is equal to:

$$N_M = 0.001 \left(F_F - F_p^* \right) A_p^{-1} v_F^{-1} \tag{6.4}$$

The result should be rounded to the lower unity. The value of N_M must always be equal or greater than 1. If this is not the case, a pump with a higher capacity must be chosen or it is also possible to fix a value for N_m and to determine the required F_F value to operate the process correctly. In any case, the value of F_F should be much higher than the value of F_p^*. This condition represents a necessary hypothesis to further calculations presented here in this handbook.

In case of batch membrane process, compared to continuous processes, the value of F_F depends only on the pump capacity and not on the available feedstock as a function of time. Batch membrane processes may be operated even if the feedstock volume is small, since the concentrate is immediately recycled back to the feed tank. Moreover, the process can work accepting discontinuous feed streams to the feed tank exhibiting high flexibility in processing different feedstock characterized by different initial KP values and washing/cleaning procedures can be integrated to the process with ease.

In case of adopting the FC control strategy, the required membrane area A' can be evaluated by the following relationship:

$$A' = F_p^* (1 - \delta_F)^{-1} J_b \left(KP_c' \right)^{-1} \tag{6.5}$$

If $J_b(KP_c') < 0$, the designed process proves to be technically not feasible. In order to consider the loss of permeability after every membrane cleaning procedure, the final required membrane area A_F, if C is the number of desired separation cycles lasting τ hours, becomes:

$$A_F = A'(1 + (C - 1)\tau \Delta w\%) \tag{6.6}$$

The set-point of the feed flow rate should be set to:

$$F_{SP} = F_p^* \tag{6.7}$$

By this approach, once a set-point value of the permeate feed flow rate controller is defined and after the correct estimation of the required membrane area to use to reach the target performances, a simple proportional integral (PI) control system can be used. Moreover, this approach guarantees operation of the membrane plant far away from the boundary condition for most of the time. Adjustments to the set-point value are needed only in case of a changing feedstock: in this case, a fingerprint in terms of KP of the feedstock should be monitored as input to an inferential advanced control system, capable of evaluating proper permeate flow rate set-point values by Eqn (2.15).

If the adopted strategy is of the PC type (control of the operating pressure), the first step is to consider the boundary TMP_b value, reduced by a safety margin δ_P of 5–10%. Starting from Eqn (2.5) it is possible to calculate the value of P_{SP}. The required membrane area A_P can be evaluated by the following relationship:

$$A_P = F_p^*\big(J_b(0)\big) - 0.5\,\alpha\,TMP_b\big(1 - \delta_p\big)\tau\big)^{-1}\big(1 + (C - 1)\tau\Delta w\%\big) \qquad (6.8)$$

In general, the set-point operating pressure value is equal to:

$$P_{SP} = TMP_b\big(1 - \delta_p\big) + \pi\big(KP\big)\big|_{t=0} \qquad (6.9)$$

By this approach, once a set-point value of the operating pressure is defined and after the correct estimation of the required membrane area to use to reach the target performances, a simple PI control system can be used. Adjustments to the set-point value are required by changes of the feed stream: in this case, a fingerprint in terms of KP of the feedstock should be monitored as input to an inferential advanced control system, capable of evaluating the osmotic pressure and the correct P_{SP} values by Eqn (2.15) at the start of operation.

It is interesting to notice that unavoidably A_P is equal or lower than A_F. Beside the ease of controlling the process by a PC strategy, due to the fact that only initial boundary flux conditions must be considered, an additional benefit are represented by reduced investment costs. On the other hand, operation will be performed constantly at a closer distance to the boundary flux conditions, equal to the fixed safety margin which should be therefore fixed at high as possible.

The number of membrane modules is defined by the area of each module, equal to A_m, by the following relationship:

$$N = AA_m^{-1} \qquad (6.10)$$

where A is equal to A_F or A_P in case of adopting an FC (control of the permeate flow rate) or PC strategy, respectively. The obtained number of membrane modules should always be rounded to the upper unity. By adoption of this procedure, selectivity and productivity are on target, while maximum longevity is guaranteed, but at this point the process is designed for all membrane modules working at the same operating conditions, and this requirement is met as long as no membrane is put in series—that is, as long as $N \leq N_M$. It is up to the discretion of the designer whether to accept this value or to try to put membrane modules in series by lowering F_F, that is, by using a pump characterized by a lower flow rate capacity.

As soon as $N > N_M$, the membrane process module layout must be studied. With N_M the number of the membrane module parallel lines connected to one pump and N_S the membrane modules in series on each line, the following equation should, at the end, hold:

$$N_S = NN_M^{-1} \qquad (6.11)$$

The value of N_S, again rounded to the upper unity, depends on the pressure drop across a single membrane module of the series ΔP_M. If a precise value is not known from the supplier, some typical values reported in Table 5.1 can be applied.

Because starting from the entrance of the first membrane module boundary flux conditions must be attained, the TMP values at the exit and in every successive membrane module are lower, thus leading to lower rejection values and permeabilities.

Two conditions need to be checked.

The first check is performed on the pressure. The following equation must be valid:

$$TMP_b - N_S \Delta P_M < 0 \tag{6.12}$$

The second check is on the rejection, calculated at the end of the membrane series, and the following condition must hold:

$$R^* < R(TMP_b - (N_S - 1)\Delta P_M) \tag{6.13}$$

If all conditions are met, the draft technical design is completed successfully.

If any of the conditions are not met, the complete procedure must be again by increasing N_S by one unit or starting from Eqn (6.1) using higher-flow-rate pumps with increased F_F values or splitting F_p^* among separated membrane systems served independently by N_P pumps. In the worst case, the considered membrane must be changed, leading to the definition of all the involved parameters again.

The optimization of economics must be evaluated by calculating the best values of N, τ, and C in order to minimize the main costs of the process $.

The guidelines presented here represent a simplified method to perform a basic economics check, used to compare different membrane processes and layouts, since some expenses are not considered, such as personal costs and extraordinary maintenance.

For this procedure, concerning the membranes, the unit membrane module cost $m and the unit housing cost $h hosting M modules must be known. Moreover, the operating life Φ of the plant must be fixed. The units of Φ must be the same as of τ.

In the first instance, concerning membranes only, the initial capital costs are equal to:

$$\$c^* = N\$m + NM^{-1}\$h \tag{6.14}$$

Once the membrane modules are depleted and need replacement, all membrane modules must be substituted. During the operating lifetime of the plant, the following hardware costs must be considered:

$$\$c' = \$c^*\left(1 + \Phi\tau^{-1}C^{-1}\right) \tag{6.15}$$

The number of pumps depends on the process layout and the required feed stream velocity v_F.

The designer need only check if the value of F_F exceeds the maximum flow rate of the pump at the required operating pressure values. If not, one pump is sufficient; in the other case, pumps with higher capacity or additional pumps in parallel are required.

If N_P is the number of adopted pumps, $cp the unit cost, and τ_P the expected lifetime, the relative investment costs can be evaluated as:

$$\$c'' = N_P \$cp \Phi \tau_P^{-1} \qquad (6.16)$$

If the pump efficiency η is fixed, normally assumed to be around 0.7, the operating costs as a function of the electricity costs $e for each kWh may be estimated as equal to:

$$\$o = 2{,}78 \ 10^{-6} \ N_P \ F_F \ \rho \ g \ TMP_b \ \Phi \ \$e \ \eta^{-1} \qquad (6.17)$$

where ρ is the feed stream density expressed in kg L^{-1} and g is equal to 9.81 m s^{-2}.

The main total costs $ are therefore:

$$\$ = \$' + \$'' + \$o \qquad (6.18)$$

Equation (6.18) requires minimization as a function of the parameters N, τ, and C by additional software, such as Excel or Memphys [18]. If the total costs are too high, most probably the project target values, such as Y^*, were set too high and requires revision.

Chapter 7

The Boundary Flux Database

Membrane process design, control, and optimization require the input of many parameters in a trustable model. The determination of the parameters is time costly and may not be performed easily.

The database represents the main core of this handbook, and the starting point for proper membrane process design for several feedstock and operating conditions. It contains all the required data to build up the model and apply the methods described in detail in the previous chapters of this handbook.

The information here reported is mostly taken from scientific, peer-reviewed, and reliable journal papers during the last decade. Only complete data sheets with all the elements required for proper process design were reported.

One of the main problems encountered in reporting data was the overall lack of a value for the parameter $\Delta w\%$. Without this parameter, proper design is not possible, but in sub-boundary flux conditions the value of this parameter is normally very low, estimated to be in the range of $0.005-0.02\%$ h^{-1}, that is equal to $5 \times 10^{-5} - 2 \times 10^{-4}$. The lower range applies well to critical, the upper range to threshold flux conditions.

The feedstock substances and/or names in the boundary flux database are presented in alphabetical order in Tables 7.1–7.95.

In the following tables, the boundary flux types were divided into critical ("C") and threshold ("TH") flux ones.

The database and the guideline given in the previous chapters of this book serve the purpose of a guideline, that may aid, but not completely substitute the experience and the know-how of the designer, and such as, should always be checked on lab scale before scaling up the processes to plant scale.

The Boundary Flux Handbook. http://dx.doi.org/10.1016/B978-0-12-801589-6.00007-3

TABLE 7.1 Agricultural Drainage Water

Reference ID		[54]
Feedstock	Key parameter	TDS
	Value in feed stream	7999 mg L^{-1}
	Pretreatments	Prefiltered by 5-μm gradient density polypropylene and 0.2 nylon cartridges
Membrane properties	Membrane type	RO
	Membrane model	Plate and frame aromatic PA TFC
	Membrane ID	LFC1
	Membrane supplier	Hydranautics
	Pore size	N/A
	w (L h^{-1} m^{-2} bar^{-1})	85.0
Process properties	T (°C)	20.0 ± 0.5
	v_F (m s^{-1})	0.11
	π (bar)	3.0
	R (%)	98.0
Boundary flux data	Boundary flux type	TH
	α (L h^{-2} m^{-2} bar^{-1})	0.0548
	Δw% (%)	N/A
	J_b (L h^{-1} m^{-2})	34.3
	TMP$_b$ (bar)	25.0

TABLE 7.2 Aldrin Solution

Reference ID		[55]		
Feedstock	Key parameter	Aldrin concentration		
	Value in feed stream	$10\ \mu g\ L^{-1}$		
	Pretreatments	None		
Membrane properties	Membrane type	NF		
	Membrane model	Aromatic polyamide	PES	Flat sheet TFC
	Membrane ID	Desal51HL	N30F	NF270
	Membrane supplier	GE Osmonics	Dow FilmTec™	Dow FilmTec™
	Pore size	300 Da	580 Da	180 Da
	$w\ (L\ h^{-1}\ m^{-2}\ bar^{-1})$	15.29 ± 0.92	9.78 ± 1.35	14.31 ± 1.75
Process properties	$T\ (^{\circ}C)$	25.0		
	$F_F\ (L\ h^{-1})$	5000		
	$\pi\ (bar)$	0.058		
	$R\ (\%)$	97.1	94.3	96.4
Boundary flux data	Boundary flux type	C	C	C
	$\alpha\ (L\ h^{-2}\ m^{-2}\ bar^{-1})$	–	–	–
	$\Delta w\%\ (\%)$	–	–	–
	$J_b\ (L\ h^{-1}\ m^{-2})$	84.0	92.0	45.0
	$TMP_b\ (bar)$	8.0	8.0	8.0

TABLE 7.3 Algae Culture of *Chlorella sorokiniana*

Reference ID		[23]
Feedstock	Key parameter	Algae concentration
	Value in feed stream	29 mg L^{-1}
	Pretreatments	None
Membrane properties	Membrane type	MF
	Membrane model	Flat sheet
	Membrane ID	Anodisc
	Membrane supplier	Whatman
	Pore size	200 nm
	w (L h^{-1} m^{-2} bar^{-1})	0.2
Process properties	T (°C)	22.0
	v$_F$ (m s^{-1})	0.24
	π (bar)	0.0
	R (%)	99.9
Boundary flux data	Boundary flux type	TH
	α (L h^{-2} m^{-2} bar^{-1})	150
	Δw% (%)	N/A
	J$_b$ (L h^{-1} m^{-2})	104.0
	TMP$_b$ (bar)	0.1

TABLE 7.4 Algae Culture of *Microcystis aeruginosa*

Reference ID		[56]
Feedstock	Key parameter	COD
	Value in feed stream	5 mg L^{-1}
	Pretreatments	500 nm sieving
Membrane properties	Membrane type	MF
	Membrane model	Tubular ceramic
	Membrane ID	Ceram
	Membrane supplier	Tami
	Pore size	100 nm
	w (L h^{-1} m^{-2} bar^{-1})	140.0
Process properties	T (°C)	22.0
	v$_F$ (m s^{-1})	N/A
	π (bar)	0.0
	R (%)	35.0
Boundary flux data	Boundary flux type	TH
	α (L h^{-2} m^{-2} bar^{-1})	2.0
	Δw% (%)	N/A
	J$_b$ (L h^{-1} m^{-2})	46.2
	TMP$_b$ (bar)	0.7

TABLE 7.5 Ammonium Solution

	Key parameter				
Reference ID		[57]			
Feedstock	Value in feed stream	6.5 ± 0.5 mg L^{-1}			
	Pretreatments	None			
Membrane properties	Membrane type	NF			
	Membrane model	Flat-sheet thin-film composite polyamide/polysulfone	Flat-sheet semiaromatic piperazine-based polyamide/polysulfone	Flat-sheet semiaromatic piperazine-based polyamide/polysulfone	Flat-sheet polyamide thin-film composite
	Membrane ID	NF90	NF200	NF270	DSHR98PP
	Membrane supplier	Dow Liquid Separations			Alfa Laval
	Pore size	200 kDa	300 kDa	580 kDa	N/A
	w (L h^{-1} m^{-2} bar^{-1})	9.8	59.0	14.3	57.6
Process properties	T (°C)	15.0			
	v_F (m s^{-1})	N/A			
	π (bar)	0.042			
	R (%)	100.0	98.4	76.8	99.7
Boundary flux data	Boundary flux type	C	C	C	C
	α (L h^{-2} m^{-2} bar^{-1})	–	–	–	–
	Δw% (%)	–	–	–	–
	J_b (L h^{-1} m^{-2})	35.6	42.0	81.3	36.7
	TMP_b (bar)	16.0	16.0	16.0	24.5

TABLE 7.6 Anaerobic MBR Sludge Solution

Reference ID		[58]	
Feedstock	Key parameter	TSS	
	Value in feed stream	6 g L^{-1}	18 g L^{-1}
	Pretreatments	None	None
Membrane properties	Membrane type	UF	
	Membrane model	Tubular	
	Membrane ID	10-HFM-276-PVI	
	Membrane supplier	ABCOR-FEG	
	Pore size	N/A	N/A
	w (L h^{-1} m^{-2} bar^{-1})	N/A	N/A
Process properties	T (°C)	20	
	v_F (m s^{-1})	N/A	N/A
	π (bar)	0.0	0.0
	R (%)	94.0	94.0
Boundary flux data	Boundary flux type	TH	TH
	α (L h^{-2} m^{-2} bar^{-1})	152.0	780.0
	Δw% (%)	N/A	N/A
	J_b (L h^{-1} m^{-2})	34.0	8.0
	TMP$_b$ (bar)	0.05	0.01

TABLE 7.7 Aniline Solution

Reference ID		[59]		
Feedstock	Key parameter	Aniline concentration		
	Value in feed stream	5 ppm		
	Pretreatments	SDS addition		
Membrane properties	Membrane type	Micellar-enhanced UF	NF	UF
	Membrane model	Polysulfone	PES	Polysulfone
	Membrane ID	–	NP010	UFX5
	Membrane supplier	Lab Made	Microdyn® Nadir	Alfa Laval
	Pore size	5 kDa	1 kDa	5 kDa
	w (L h^{-1} m^{-2} bar^{-1})	65.0	16.0	60.0
Process properties	T (°C)	25.0		
	v_F (m s^{-1})	N/A	N/A	N/A
	π (bar)	0.0	0.0	0.0
	R (%)	26.0	50.0	23.0
Boundary flux data	Boundary flux type	C	C	C
	α (L h^{-2} m^{-2} bar^{-1})	–	–	–
	Δw% (%)	–	–	–
	J_b (L h^{-1} m^{-2})	44.6	8.3	29.3
	TMP_b (bar)	2.5	10.0	2.5

TABLE 7.8 Anionic Surfactant Solution

Reference ID		[60]
Feedstock	Key parameter	CL80 anionic surfactant
	Value in feed stream	0.5–1.0 g L^{-1}
	Pretreatments	None
Membrane properties	Membrane type	NF
	Membrane model	Flat-sheet TFC PA/PS
	Membrane ID	DL
	Membrane supplier	Uwatech Gmbh
	Pore size	N/A
	w (L h^{-1} m^{-2} bar^{-1})	21.0
Process properties	T (°C)	20.0
	v_F (m s^{-1})	N/A
	π (bar)	0.05
	R (%)	94.0
Boundary flux data	Boundary flux type	TH
	α (L h^{-2} m^{-2} bar^{-1})	0.6375
	Δw% (%)	N/A
	J_b (L h^{-1} m^{-2})	9.2
	TMP$_b$ (bar)	20.0

TABLE 7.9 Arsenic Solution

Reference ID	Key parameter	[61]				
Feedstock	Key parameter	Arsenic concentration				
	Value in feed stream	49.5 µg L⁻¹				
	Pretreatments	None				
Membrane properties	Membrane type	RO				
	Membrane model	Spiral wound polymeric TFC PA/PS				
	Membrane ID	SW30HR	BW30LE	SCW5	ESPAB	ESPA2
	Membrane supplier	Filmtec		Hydranautics		
	Pore size	–	–	–	–	–
	w (L h⁻¹ m⁻² bar⁻¹)	0.78	2.14	1.16	2.57	2.96
Process properties	T (°C)	20.0				
	F_F (L h⁻¹)	120.0				
	π (bar)	12.2				
	R (%)	99.0	91.0	99.0	95.0	86.0
Boundary flux data	Boundary flux type	C	C	C	C	C
	α (L h⁻² m⁻² bar⁻¹)	–	–	–	–	–
	$\Delta w\%$ (%)	N/A	N/A	N/A	N/A	N/A
	J_b (L h⁻¹ m⁻²)	31.2	85.6	46.4	102.0	118.4
	TMP_b (bar)	24.0	30.0	24.0	40.0	24.0

TABLE 7.10 Atrazine Solution

Reference ID		[62]				
Feedstock	Key parameter	Atrazine concentration				
	Value in feed stream	4.1 mg L^{-1}				
	Pretreatments	None				
Membrane properties	Membrane type	NF	RO	RO	NF	NF
	Membrane model	Polyamide		Polypiperazine		
	Membrane ID	NF90	XLE	BW30	NF99	NF99HF
	Membrane supplier	Dow Chemicals		Alfa Laval		
	Pore size	0.81 nm	–	–	0.85 nm	0.82 nm
	w (L h^{-1} m^{-2} bar^{-1})	36.9	38.3	19.9	44.3	60.4
Process properties	T (°C)	25.0				
	v_F (m s^{-1})	0.133				
	π (bar)	0.41				
	R (%)	93.2	88.7	97.7	98.2	98.2
Boundary flux data	Boundary flux type	C	C	C	C	C
	α (L h^{-2} m^{-2} bar^{-1})	–	–	–	–	–
	Δw% (%)	–	–	–	–	–
	J_b (L h^{-1} m^{-2})	–	–	–	–	–
	TMP$_b$ (bar)	10.0	10.0	10.0	10.0	10.0

TABLE 7.11 Baker Yeast Wastewater

Reference ID		[63]	
Feedstock	Key parameter	Yeast concentration	
	Value in feed stream	1.00%	
	Pretreatments	pH adjustment to 4.0	
Membrane properties	Membrane type	UF	UF
	Membrane model	Flat sheet	Flat sheet
	Membrane ID	GR51	C30G
	Membrane supplier	Dow	Hoechst Company
	Pore size	30 kDa	30 kDa
	w (L h^{-1} m^{-2} bar^{-1})	1801.0	520.0
Process properties	T (°C)	25.0	
	v$_F$ (m s^{-1})	0.19	0.19
	π (bar)	0.0	0.0
	R (%)	70.0	70.0
Boundary flux data	Boundary flux type	C	C
	α (L h^{-2} m^{-2} bar^{-1})	–	–
	Δw% (%)	–	–
	J$_b$ (L h^{-1} m^{-2})	19.0	16.0
	TMP$_b$ (bar)	0.17	0.17

TABLE 7.12 Barium Solution

Reference ID		[64]	
Feedstock	Key parameter	Barium concentration	
	Value in feed stream	2000 ppm	
	Pretreatments	None	
Membrane properties	Membrane type	RO	NF
	Membrane model	Flat-sheet aromatic PA/PS TFC	Flat-sheet PES
	Membrane ID	BW30	NF1
	Membrane supplier	Amei Ande Membrane Technology Ltd	
	Pore size	–	N/A
	w (L h^{-1} m^{-2} bar^{-1})	3.4	7.3
Process properties	T (°C)	25.0	
	F_F (L h^{-1})	400.0	
	π (bar)	1.64	
	R (%)	85.4	25.3
Boundary flux data	Boundary flux type	C	C
	α (L h^{-2} m^{-2} bar^{-1})	–	–
	$\Delta w\%$ (%)	–	–
	J_b (L h^{-1} m^{-2})	60.0	180.0
	TMP_b (bar)	6.0	6.0

TABLE 7.13 Beverage Production Wastewater

Reference ID		[65]
Feedstock	Key parameter	COD
	Value in feed stream	3.75 g L^{-1}
	Pretreatments	None
Membrane properties	Membrane type	UF
	Membrane model	TiO$_2$ ceramic single-channel
	Membrane ID	Membralox T1-70
	Membrane supplier	Pall Corporation
	Pore size	5 nm
	w (L h^{-1} m^{-2} bar^{-1})	27.0
Process properties	T (°C)	20.0
	v$_f$ (m s^{-1})	3.1
	π (bar)	0.0
	R (%)	27.5
Boundary flux data	Boundary flux type	TH
	α (L h^{-2} m^{-2} bar^{-1})	0.33
	Δw% (%)	N/A
	J$_b$ (L h^{-1} m^{-2})	342.2
	TMP$_b$ (bar)	3.0

TABLE 7.14 Boric Acid Solution

	Key parameter								
Reference ID		[66]							
Feedstock		CaSO$_4$, humic acid, silica colloid, sodium alginate concentration							
	Value in feed stream	1 g L^{-1}, 20, 20, 20 mg L^{-1}							
	Pretreatments	None							
Membrane properties	Membrane type	NF	RO	NF	RO	NF	RO	NF	RO
	Membrane model	Flat-sheet fully aromatic PA/PS TFC							
	Membrane ID	NF270	BW30	NF270	BW30	NF270	BW30	NF270	BW30
	Membrane supplier	Dow FilmTec™							
	Pore size	0.84 nm	–	0.84 nm	–	0.84 nm	–	0.84 nm	–
	w (L h^{-1} m^{-2} bar^{-1})	3.5	14.0	3.5	14.0	3.5	14.0	3.5	14.0
Process properties	T (°C)	20.0 ± 0.1							
	v$_F$ (m s^{-1})	0.3							
	π (bar)	0.426		0.418		0.418		0.418	
	R (%)	35.8	96.2	35.8	96.2	35.8	96.2	35.8	96.2
Boundary flux data	Boundary flux type	TH	TH	TH	TH	C	TH	TH	TH
	α (L h^{-2} m^{-2} bar^{-1})	0.42	0.32	2.97	0.35	–	0.22	1.98	0.27
	Δw% (%)	N/A	N/A	N/A	N/A	N/A	N/A	N/A	N/A
	J$_b$ (L h^{-1} m^{-2})	75.6	40.8	29.9	43.2	58.8	47.7	38.6	39.9
	TMP$_b$ (bar)	6.0	17.0	6.0	17.0	6.0	17.0	6.0	17.0

TABLE 7.15 Boron Solution

Reference ID	Key parameter	[67]			
Feedstock	Value in feed stream	Boron concentration			
		5.1 mg L^{-1}			
	Pretreatments	None		1 μm cartridge filters	
Membrane properties	Membrane type	RO			
	Membrane model	Flat-sheet TFC PA/PS			
	Membrane ID	UTC-80-AB	SW30HR	UTC-80-AB	SW30HR
	Membrane supplier	Toray™	Filmtec™	Toray™	Filmtec™
	Pore size	–	–	–	–
	w (L h^{-1} m^{-2} bar^{-1})	1.6	0.53	1.25	0.44
Process properties	T (°C)	22.0			
	v_F (m s^{-1})	0.9			
	π (bar)	0.05			
	R (%)	98.8	92.9	89.0	86.7
Boundary flux data	Boundary flux type	C	TH	C	TH
	α (L h^{-2} m^{-2} bar^{-1})	–	0.0091	–	0.0032
	Δw% (%)	–	N/A	–	N/A
	J_b (L h^{-1} m^{-2})	76.4	16.9	60.5	17.7
	TMP$_b$ (bar)	48.2	48.2	48.2	48.2

TABLE 7.16 Bovine Serum Albumin (BSA)

Reference ID		[68]	
Feedstock	Key parameter	Protein (PN) on polysaccharide (PS) ratio	
	Value in feed stream	1000/50 mg L^{-1}	
	Pretreatments	0.45 μm filter	
Membrane properties	Membrane type	MF	
	Membrane model	Spiral wound polyvinylidene fluoride (PVDF)	Spiral wound polycarbonate (PC)
	Membrane ID	GVWP 025 00	GTTP 025 00
	Membrane supplier	Millipore Corp.	
	Pore size	220 nm	220 nm
	w (L h^{-1} m^{-2} bar^{-1})	6000	
Process properties	T (°C)	25.0	
	F$_F$ (L h^{-1})	500.0	
	π (bar)	0.0	
	R (%)	PN 1.0; PS 11.9	PN 0.6; PS 3.6
Boundary flux data	Boundary flux type	TH	TH
	α (L h^{-2} m^{-2} bar^{-1})	3672.0	175.0
	Δw% (%)	N/A	N/A
	J$_b$ (L h^{-1} m^{-2})	352.8	108.9
	TMP$_b$ (bar)	0.5	0.5

TABLE 7.17 Brackish Water

	Key parameter	Reference ID	[69]
Feedstock	Salt concentration		
	Value in feed stream	2 g L^{-1}	
	Pretreatments	None	
Membrane properties	Membrane type	RO	
	Membrane model	Flat-sheet PS modified	Flat-sheet TFC
	Membrane ID	N/A	FT 30
	Membrane supplier	Lab Made	Dow Water
	Pore size	–	–
	w (L h^{-1} m^{-2} bar^{-1})	N/A	N/A
Process properties	T (°C)	25.0	
	v_F (m s^{-1})	N/A	
	π (bar)	3.03	
	R (%)	98.4	99.0
Boundary flux data	Boundary flux type	C	C
	α (L h^{-2} m^{-2} bar^{-1})	–	–
	Δw% (%)	–	–
	J$_b$ (L h^{-1} m^{-2})	102.5	41.6
	TMP$_b$ (bar)	15.5	15.5

The Boundary Flux Database Chapter | 7 **59**

TABLE 7.18 Buffered Synthetic Water Solution

Reference ID		[70]	
Feedstock	Key parameter	Sodium acetate trihydrate, active cells concentration (*Pseudomonas fluorescens* Migula)	
	Value in feed stream	130.14 ppm, 0 cells mL^{-1}	130.14 ppm, 103 cells mL^{-1}
	Pretreatments	None	
Membrane properties	Membrane type	UF	UF
	Membrane model	Flat-sheet cellulose acetate	
	Membrane ID	N/A	N/A
	Membrane supplier	GE Water and Process Technologies	
	Pore size	20 kDa	
	w (L h^{-1} m^{-2} bar^{-1})	N/A	
Process properties	T (°C)	26.0	
	v_f (m s^{-1})	N/A	
	π (bar)	0.0	
	R (%)	N/A	
Boundary flux data	Boundary flux type	TH	TH
	α (L h^{-2} m^{-2} bar^{-1})	0.81	2.87
	Δw% (%)	N/A	N/A
	J_b (L h^{-1} m^{-2})	31.2	29.4
	TMP$_b$ (bar)	1.72	1.72

TABLE 7.19 Cadmium Solution

Reference ID		[71]
Feedstock	Key parameter	Cadmium concentration
	Value in feed stream	0.35 ppm
	Pretreatments	Surfactant addition (SDS)
Membrane properties	Membrane type	Micellar-enhanced UF
	Membrane model	Spiral wound polyethersulfone
	Membrane ID	PW series
	Membrane supplier	GE Osmonics Labstore
	Pore size	10 kDa
	w (L h^{-1} m^{-2} bar^{-1})	42.27
Process properties	T (°C)	21.0
	v_F (m s^{-1})	0.24
	π (bar)	0.082
	R (%)	85.0
Boundary flux data	Boundary flux type	C
	α (L h^{-2} m^{-2} bar^{-1})	–
	$\Delta w\%$ (%)	–
	J_b (L h^{-1} m^{-2})	124.4
	TMP$_b$ (bar)	3.0

TABLE 7.20 Car Manufacturer Wastewater

Reference ID		[72]
Feedstock	Key parameter	TDS
	Value in feed stream	79.9 mg L^{-1}
	Pretreatments	None
Membrane properties	Membrane type	UF
	Membrane model	Single channel ceramic ZrO$_2$
	Membrane ID	T1-70
	Membrane supplier	Membralox
	Pore size	50 nm
	w (L h^{-1} m^{-2} bar^{-1})	906.0
Process properties	T (°C)	24.0
	v$_F$ (m s^{-1})	3.2
	π (bar)	0.0
	R (%)	99.95
Boundary flux data	Boundary flux type	TH
	α (L h^{-2} m^{-2} bar^{-1})	9.78
	Δw% (%)	N/A
	J$_b$ (L h^{-1} m^{-2})	97.2
	TMP$_b$ (bar)	1.0

TABLE 7.21 Chromate Solution

Reference ID		[73]	
Feedstock	Key parameter	Molar ratio chromate to cetylpyridinium chloride	
	Value in feed stream	20/100 mg L^{-1}	20/200 mg L^{-1}
	Pretreatments	Activated carbon fiber 10 μm	
Membrane properties	Membrane type	Micellar-enhanced UF	
	Membrane model	Hollow fiber polyacrylonitrile	
	Membrane ID	N/A	
	Membrane supplier	N/A	
	Pore size	100 kDa	
	w (L h^{-1} m^{-2} bar^{-1})	31.2	
Process properties	T (°C)	N/A	
	F$_F$ (L h^{-1})	6.0	
	π (bar)	0.0	
	R (%)	98.6	
Boundary flux data	Boundary flux type	TH	TH
	α (L h^{-2} m^{-2} bar^{-1})	33.1	45.6
	Δw% (%)	–	–
	J$_b$ (L h^{-1} m^{-2})	33.3	24.6
	TMP$_b$ (bar)	1.4	1.4

TABLE 7.22 Chromium Solution

Reference ID		[74]
Feedstock	Key parameter	Chromate concentration
	Value in feed stream	0.2 mM
	Pretreatments	Cetyltrimethylammoniumbromide (CTAB) 4 mM
Membrane properties	Membrane type	MF
	Membrane model	Flat-sheet anisotropic cellulose acetate
	Membrane ID	N/A
	Membrane supplier	Schleicher & Schuell
	Pore size	200 nm
	w (L h^{-1} m^{-2} bar^{-1})	7000.0
Process properties	T (°C)	30.0
	v_F (m s^{-1})	6.0
	π (bar)	0.0
	R (%)	75.0
Boundary flux data	Boundary flux type	TH
	α (L h^{-2} m^{-2} bar^{-1})	1.16
	$\Delta w\%$ (%)	–
	J_b (L h^{-1} m^{-2})	239.5
	TMP_b (bar)	1.5

TABLE 7.23 Cocoon Wastewater

Reference ID		[75]
Feedstock	Key parameter	Serecin concentration
	Value in feed stream	6508 mg L^{-1}
	Pretreatments	pH adjustment to 4.5
Membrane properties	Membrane type	NF
	Membrane model	Flat sheet
	Membrane ID	DK
	Membrane supplier	GE Osmonics
	Pore size	2 nm
	w (L h^{-1} m^{-2} bar^{-1})	N/A
Process properties	T (°C)	18.0
	v$_F$ (m s^{-1})	N/A
	π (bar)	0.043
	R (%)	86.1
Boundary flux data	Boundary flux type	C
	α (L h^{-2} m^{-2} bar^{-1})	–
	Δw% (%)	–
	J$_b$ (L h^{-1} m^{-2})	2.6
	TMP$_b$ (bar)	4.9

TABLE 7.24 Coking Wastewater

Reference ID		[76]	
Feedstock	Key parameter	Ammonium salts concentration	NF diafiltration
	Value in feed stream	60 g L^{-1}	
	Pretreatments	Zirconia membrane (50 nm) prefiltration	
Membrane properties	Membrane type	NF continuous	
	Membrane model	Spiral wound TFC PA/PS	
	Membrane ID	DK1812	
	Membrane supplier	GE Corp.	
	Pore size	0.5 nm	
	w (L h^{-1} m^{-2} bar^{-1})	7.6	
Process properties	T (°C)	30.0	
	F$_F$ (L h^{-1})	780.0	
	π (bar)	N/A	
	R (%)	95.0	93.4
Boundary flux data	Boundary flux type	TH	TH
	α (L h^{-2} m^{-2} bar^{-1})	2.19	5.0
	Δw% (%)	N/A	N/A
	J$_b$ (L h^{-1} m^{-2})	29.2	18.5
	TMP$_b$ (bar)	20.0	20.0

TABLE 7.25 Copper Solution

Reference ID		[71]
Feedstock	Key parameter	Copper concentration
	Value in feed stream	0.4 mg L^{-1}
	Pretreatments	Surfactant addition (SDS)
Membrane properties	Membrane type	Micellar-enhanced UF
	Membrane model	Spiral wound polyethersulfone
	Membrane ID	PW series
	Membrane supplier	GE Osmonics Labstore
	Pore size	10 kDa
	w (L h^{-1} m^{-2} bar^{-1})	42.27
Process properties	T (°C)	21.0
	v$_F$ (m s^{-1})	0.24
	π (bar)	0.0
	R (%)	81.0
Boundary flux data	Boundary flux type	C
	α (L h^{-2} m^{-2} bar^{-1})	–
	Δw% (%)	–
	J$_b$ (L h^{-1} m^{-2})	169.1
	TMP$_b$ (bar)	3.0

TABLE 7.26 Dairy Fluids

Reference ID		[77]	
Feedstock	Key parameter	Whey concentration	
	Value in feed stream	10 g L^{-1}	
	Pretreatments	None	
Membrane properties	Membrane type	MF	UF
	Membrane model	Monotubular ceramic membranes α-alumina on α-alumina support	Monotubular ceramic membranes ZrO$_2$ on α-alumina support
	Membrane ID	–	–
	Membrane supplier	SCT	
	Pore size	200 nm	50 nm
	w (L h^{-1} m^{-2} bar^{-1})	687.5	1634.5
Process properties	T (°C)	25.0	
	v_F (m s^{-1})	0.43	
	π (bar)	0.0	
	R (%)	N/A	N/A
Boundary flux data	Boundary flux type	TH	TH
	α (L h^{-2} m^{-2} bar^{-1})	27.3	12.0
	Δw% (%)	–	–
	J_b (L h^{-1} m^{-2})	21.5	22.0
	TMP$_b$ (bar)	0.3	0.3

TABLE 7.27 Dairy UF Permeate

Reference ID		[78]
Feedstock	Key parameter	KCl, CaCl$_2$, KH$_2$PO$_4$, KH concentration
	Value in feed stream	10, 2, 2, 1 nM
	Pretreatments	UF 10 kDa
Membrane properties	Membrane type	NF
	Membrane model	Flat-sheet PA/PS TFC
	Membrane ID	TFC-SR3
	Membrane supplier	Koch Membrane Systems
	Pore size	1 nm
	w (L h^{-1} m^{-2} bar^{-1})	6.4
Process properties	T (°C)	16.0
	v$_F$ (m s^{-1})	0.45
	π (bar)	0.07
	R (%)	40.0
Boundary flux data	Boundary flux type	TH
	α (L h^{-2} m^{-2} bar^{-1})	1.55
	Δw% (%)	–
	J$_b$ (L h^{-1} m^{-2})	81.1
	TMP$_b$ (bar)	15.0

TABLE 7.28 Dairy Wastewater

Reference ID		[79]
Feedstock	Key parameter	COD
	Value in feed stream	36,000 mg L^{-1}
	Pretreatments	Two sieves with pore size of 0.25 and 0.10 mm
Membrane properties	Membrane type	NF
	Membrane model	Rotating polyamide
	Membrane ID	NF270
	Membrane supplier	Dow Filmtec
	Pore size	200 Da
	w (L h^{-1} m^{-2} bar^{-1})	11.3
Process properties	T (°C)	35.0
	v$_f$ (rpm)	2000.0
	π (bar)	1.0
	R (%)	99.85
Boundary flux data	Boundary flux type	TH
	α (L h^{-2} m^{-2} bar^{-1})	0.645
	Δw% (%)	N/A
	J$_b$ (L h^{-1} m^{-2})	319.35
	TMP$_b$ (bar)	40.0

TABLE 7.29 Dye Solution

Reference ID		[80]
Feedstock	Key parameter	Dye concentration
	Value in feed stream	1000 mg L^{-1}
	Pretreatments	None
Membrane properties	Membrane type	NF
	Membrane model	Aromatic poly (*m*-phenylene isophthalamide) with polyvinyl pyrrolidone and LiCl additives
	Membrane ID	–
	Membrane supplier	Lab Made
	Pore size	416 Da
	w (L h^{-1} m^{-2} bar^{-1})	N/A
Process properties	T (°C)	20.0
	v$_F$ (m s^{-1})	N/A
	π (bar)	0.042
	R (%)	99.4
Boundary flux data	Boundary flux type	TH
	α (L h^{-2} m^{-2} bar^{-1})	2.9
	Δw% (%)	N/A
	J$_b$ (L h^{-1} m^{-2})	64.6
	TMP$_b$ (bar)	6.0

TABLE 7.30 *Escherichia coli*

Reference ID		[81]	
Feedstock	Key parameter	*E. Coli* concentration	
	Value in feed stream	10^6 cfu mL^{-1}	
	Pretreatments	None	Thermosonication
Membrane properties	Membrane type	RO	
	Membrane model	Flat-sheet TFC PA/PS	
	Membrane ID	UTC-70UB	
	Membrane supplier	Toray	
	Pore size	—	
	w (L h^{-1} m^{-2} bar^{-1})	0.303	
Process properties	T (°C)	25.0	
	v_F (m s^{-1})	N/A	
	π (bar)	N/A	
	R (%)	N/A	N/A
Boundary flux data	Boundary flux type	TH	TH
	α (L h^{-2} m^{-2} bar^{-1})	0.20	0.16
	Δw% (%)	N/A	N/A
	J_b (L h^{-1} m^{-2})	0.3	0.15
	TMP$_b$ (bar)	4.5	4.5

TABLE 7.31 Eutrophic Source Water with Green Algae A (*Chlorella vulgaris*)

Reference ID		[82]
Feedstock	Key parameter	Algae suspension concentration (NPDOC)
	Value in feed stream	1.79 mg L^{-1}
	Pretreatments	None
Membrane properties	Membrane type	UF
	Membrane model	Flat-sheet regenerated cellulose
	Membrane ID	8200
	Membrane supplier	Amicon Corp.
	Pore size	63.5 nm
	w (L h^{-1} m^{-2} bar^{-1})	58.82
Process properties	T (°C)	23.0
	v_f (m s^{-1})	N/A
	π (bar)	0.0
	R (%)	N/A
Boundary flux data	Boundary flux type	C
	α (L h^{-2} m^{-2} bar^{-1})	–
	Δw% (%)	–
	J_b (L h^{-1} m^{-2})	70.0
	TMP$_b$ (bar)	1.4

TABLE 7.32 Eutrophic Source Water with Green Algae B (*Chodatella sp.*)

Reference ID		[82]
Feedstock	Key parameter	Algae suspension concentration (NPDOC)
	Value in feed stream	1 mg L^{-1}
	Pretreatments	None
Membrane properties	Membrane type	UF
	Membrane model	Flat-sheet regenerated cellulose
	Membrane ID	8200
	Membrane supplier	Amicon Corp.
	Pore size	63.5 nm
	w (L h^{-1} m^{-2} bar^{-1})	58.82
Process properties	T (°C)	23.0
	v$_F$ (m s^{-1})	N/A
	π (bar)	0.0
	R (%)	N/A
Boundary flux data	Boundary flux type	C
	α (L h^{-2} m^{-2} bar^{-1})	–
	Δw% (%)	–
	J$_b$ (L h^{-1} m^{-2})	105.0
	TMP$_b$ (bar)	2.0

TABLE 7.33 Eutrophic Source Water with Green Algae C (*Microcystis sp.*)

Reference ID		[82]
Feedstock	Key parameter	Algae suspension concentration (NPDOC)
	Value in feed stream	2.2 mg L^{-1}
	Pretreatments	None
Membrane properties	Membrane type	UF
	Membrane model	Flat-sheet regenerated cellulose
	Membrane ID	8200
	Membrane supplier	Amicon Corp.
	Pore size	63.5 nm
	w (L h^{-1} m^{-2} bar^{-1})	58.82
Process properties	T (°C)	23.0
	v_F (m s^{-1})	N/A
	π (bar)	0.0
	R (%)	N/A
Boundary flux data	Boundary flux type	C
	α (L h^{-2} m^{-2} bar^{-1})	–
	Δw% (%)	–
	J_b (L h^{-1} m^{-2})	55.0
	TMP$_b$ (bar)	1.4

TABLE 7.34 Expanded Polystyrene Solution (EPS)

Reference ID		[83]			
Feedstock	Key parameter	Sodium alginate concentration (polysaccharide 200 nm)			
	Value in feed stream	80 mg L^{-1}			
Membrane properties	Pretreatments	None	20 mg L^{-1} PVA	None	20 mg L^{-1} PVA
	Membrane type	MF	MF	UF	UF
	Membrane model	Flat-sheet PES (polyethersulfone)			
	Membrane ID	—	—	—	—
	Membrane supplier	Synder			
	Pore size	200 nm	200 nm	1 kDa	1 kDa
	w (L h^{-1} m^{-2} bar^{-1})	N/A	N/A	N/A	N/A
Process properties	T (°C)	20.0			
	v$_F$ (m s^{-1})	N/A			
	π (bar)	0.0			
	R (%)	95.0			
Boundary flux data	Boundary flux type	TH	TH	TH	TH
	α (L h^{-2} m^{-2} bar^{-1})	8.44	6.44	2.4	2.0
	Δw% (%)	N/A	N/A	N/A	N/A
	J$_b$ (L h^{-1} m^{-2})	7.3	5.58	7.55	5.93
	TMP$_b$ (bar)	2.5	2.5	2.5	2.5

TABLE 7.35 Fecal Coliform Suspension

Reference ID		[84]
Feedstock	Key parameter	*Aspergillus niger, Escherichia Coli.*
	Value in feed stream	1000, 3000 mg mL^{-1}
	Pretreatments	None
Membrane properties	Membrane type	NF
	Membrane model	Nanocomposite with 1,3,4-oxadiazole (APDSPO)
	Membrane ID	PO-6
	Membrane supplier	Self-assembled
	Pore size	1 nm
	w (L h^{-1} m^{-2} bar^{-1})	5.8
Process properties	T (°C)	25.0
	v$_F$ (m s^{-1})	300.0
	π (bar)	4.5
	R (%)	100.0
Boundary flux data	Boundary flux type	TH
	α (L h^{-2} m^{-2} bar^{-1})	0.23
	Δw% (%)	N/A
	J$_b$ (L h^{-1} m^{-2})	35.2
	TMP$_b$ (bar)	6.2

TABLE 7.36 Flouride Solution

Reference ID	Key parameter	[85]		
Feedstock		Fluoride concentration		
	Value in feed stream	20 mg L^{-1}		
	Pretreatments	None		
Membrane properties	Membrane type	NF		
	Membrane model	Flat-sheet cross-flow polyamide		
	Membrane ID	NF-1	NF-2	NF-20
	Membrane supplier	N/A	N/A	N/A
	Pore size	0.53 nm	0.57 nm	0.54 nm
	w (L h^{-1} m^{-2} bar^{-1})	110	130	120
Process properties	T (°C)	35.0		
	F$_F$ (L h^{-1})	750.0		
	π (bar)	0.4		
	R (%)	95.0	78.0	86.0
Boundary flux data	Boundary flux type	TH	TH	TH
	α (L h^{-2} m^{-2} bar^{-1})	0.21	0.39	0.42
	Δw% (%)	N/A	N/A	N/A
	J$_b$ (L h^{-1} m^{-2})	131.0	317.0	174.0
	TMP$_b$ (bar)	14.0	14.0	14.0

TABLE 7.37 Gelatin Suspension

Reference ID		[86]
Feedstock	Key parameter	Gelatin concentration
	Value in feed stream	$0.01\ g\,L^{-1}$
	Pretreatments	None
Membrane properties	Membrane type	UF
	Membrane model	Spiral wound polysulfone
	Membrane ID	N/A
	Membrane supplier	Dow FilmTec
	Pore size	30 kDa
	$w\ (L\,h^{-1}\,m^{-2}\,bar^{-1})$	N/A
Process properties	$T\ (^{\circ}C)$	30.0
	$v_F\ (m\,s^{-1})$	1.0
	$\pi\ (bar)$	0.0
	$R\ (\%)$	99.9
Boundary flux data	Boundary flux type	TH
	$\alpha\ (L\,h^{-2}\,m^{-2}\,bar^{-1})$	35.0
	$\Delta w\%\ (\%)$	N/A
	$J_b\ (L\,h^{-1}\,m^{-2})$	640.3
	$TMP_b\ (bar)$	2.0

TABLE 7.38 Geothermal Water

Reference ID		[87]
Feedstock	Key parameter	Boron concentration
	Value in feed stream	11.4 mg L^{-1}
	Pretreatments	Sand filter cartridges of 10 and 5 μm in series
Membrane properties	Membrane type	RO
	Membrane model	Spiral wound polymeric
	Membrane ID	Filmtec BW30-2540
	Membrane supplier	Dow Water
	Pore size	N/A
	w (L h^{-1} m^{-2} bar^{-1})	N/A
Process properties	T (°C)	25.0
	F$_F$ (L h^{-1})	200.0
	π (bar)	0.75
	R (%)	50.0
Boundary flux data	Boundary flux type	C
	α (L h^{-2} m^{-2} bar^{-1})	–
	Δw% (%)	–
	J$_b$ (L h^{-1} m^{-2})	48.0
	TMP$_b$ (bar)	15.0

TABLE 7.39 Groundwater

Reference ID		[85]		
Feedstock	Key parameter	Fluoride (0.352 nm hydrated radius) concentration		
	Value in feed stream	20 mg L^{-1}		
	Pretreatments	None		
Membrane properties	Membrane type	NF		
	Membrane model	Flat-sheet polymeric (TFC PA)		
	Membrane ID	NF-1	NF-2	NF-20
	Membrane supplier	Sepro Membranes Inc.		
	Pore size	0.53 nm	0.57 nm	0.54 nm
	w (L h^{-1} m^{-2} bar^{-1})	110	130	120
Process properties	T (°C)	35.0		
	v_F (m s^{-1})	1.16		
	π (bar)	0.41		
	R (%)	95.0	78.0	86.0
Boundary flux data	Boundary flux type	TH	TH	TH
	α (L h^{-2} m^{-2} bar^{-1})	9.15	3.63	1.86
	Δw% (%)	N/A	N/A	N/A
	J_b (L h^{-1} m^{-2})	330.0	188.0	138.0
	TMP$_b$ (bar)	14.0	14.0	14.0

TABLE 7.40 High Salinity Drilling Water

Reference ID	Key parameter	[88]		
Feedstock	Value in feed stream	TOC		
	Pretreatments	3.86 g L^{-1} (67 g L^{-1} salinity)		
		None		
Membrane properties	Membrane type	UF	NF	NF
	Membrane model	Flat-sheet PES	Flat-sheet PES	Flat-sheet TFC PA/PES
	Membrane ID	XP117	MT44	MT03
	Membrane supplier	PCI		
	Pore size	4 kDa	2 kDa	0.2 kDa
	w (L h^{-1} m^{-2} bar^{-1})	320	185	115
Process properties	T (°C)	25.0		
	v_F (m s^{-1})	2.5		
	π (bar)	54.0		
	R (%)	74.1% TOC	46.4% TOC	40.15% TOC
Boundary flux data	Boundary flux type	TH	TH	TH
	α (L h^{-2} m^{-2} bar^{-1})	0.0085	0.0060	0.0035
	Δw% (%)	N/A	N/A	N/A
	J_b (L h^{-1} m^{-2})	202.9	157.35	100
	TMP$_b$ (bar)	20.0	20.0	20.0

TABLE 7.41 Humic Acid Solution

Reference ID		[89]			
Feedstock	Key parameter	Humic acid concentration			
	Value in feed stream	15 mg L⁻¹			
	Pretreatments	None	None	None	UV-grafting modification
Membrane properties	Membrane type	NF			
	Membrane model	TFC unmodified PES		Flat-sheet asymmetric	
	Membrane ID	NFPES10	Modified 1TMBPA	NFPES10	NFPES10
	Membrane supplier	Nadir		Hoechst (Germany)	
	Pore size	1.47 nm	1.34 nm	1.06 nm	1.55 nm
	w (L h⁻¹ m⁻² bar⁻¹)	15.55	6.66	N/A	N/A
Process properties	T (°C)	24.0		25.0	
	v_F (m s⁻¹)	24.0		24.0	
	π (bar)	0.042			
	R (%)	97.0	97.0	98.5	99.9
Boundary flux data	Boundary flux type	TH	C	TH	TH
	α (L h⁻² m⁻² bar⁻¹)	5.0	–	175.0	156.0
	$\Delta w\%$ (%)	N/A	–	N/A	N/A
	J_b (L h⁻¹ m⁻²)	72.74	43.95	47.60	81.10
	TMP_b (bar)	2.0	2.0	2.0	2.0

TABLE 7.42 Iron Solution

Reference ID		[90]
Feedstock	Key parameter	Iron concentration
	Value in feed stream	100 mg L^{-1}
	Pretreatments	pH adjustment to 7.0
Membrane properties	Membrane type	UF
	Membrane model	PES
	Membrane ID	—
	Membrane supplier	Hoechst GmbH
	Pore size	50 kDa
	w (L h^{-1} m^{-2} bar^{-1})	120.0
Process properties	T (°C)	20.0
	v_F (rpm)	400.0
	π (bar)	0.07
	R (%)	—
Boundary flux data	Boundary flux type	TH
	α (L h^{-2} m^{-2} bar^{-1})	0.167
	Δw% (%)	N/A
	J_b (L h^{-1} m^{-2})	40.0
	TMP$_b$ (bar)	2.0

TABLE 7.43 Lactic Acid Solution

Reference ID	Key parameter	[91]			
Feedstock	Value in feed stream	Microorganism (broth) concentration			
		2.1 g L^{-1}			
	Pretreatments	0.2 μm vacuum microfilter drying the cake at 378 K for 24 h			
Membrane properties	Membrane type	MF			
	Membrane model	TFC PSF/CA in the presence of 10 wt% PEG as pore former in N-methyl-2-pyrrolidone (NMP)			
	Membrane ID	MF1	MF2	MF3	MF4
	Membrane supplier	Lab Made			
	Pore size	0.24 nm	0.27 nm	0.32 nm	0.38 nm
	w (L h^{-1} m^{-2} bar^{-1})	1182.7	1288.3	1361.3	1393.8
Process properties	T (°C)	25.0			
	F$_F$ (L h^{-1})	60.0			
	π (bar)	0.0			
	R (%)	N/A	N/A	N/A	N/A
Boundary flux data	Boundary flux type	TH	TH	TH	TH
	α (L h^{-2} m^{-2} bar^{-1})	128.0	88.9	33.1	10.3
	Δw% (%)	N/A	N/A	N/A	N/A
	J$_b$ (L h^{-1} m^{-2})	1179.3	1266.2	1334.5	1353.1
	TMP$_b$ (bar)	1.5	1.5	1.5	1.5

TABLE 7.44 Landfill Leachate

Reference ID		[92]					
Feedstock	Key parameter	TOC					
	Value in feed stream	635.8 mg L^{-1}		572.2 mg L^{-1}			
	Pretreatments	None		Coagulation with aluminum sulfate		Electrocoagulation	
Membrane properties	Membrane type	NF					
	Membrane model	TFC flat-sheet PA/PS					
	Membrane ID	NF 270	SR2	NF 270	SR2	NF 270	SR2
	Membrane supplier	Dow FilmTec	Koch	Dow FilmTec	Koch	Dow FilmTec	Koch
	Pore size	0.84 nm	1.28 nm	0.84 nm	1.28 nm	0.84 nm	1.28 nm
	w (L h^{-1} m^{-2} bar^{-1})	13.5	15.4	13.5	15.4	13.5	15.4
Process properties	T (°C)	25.0					
	v_F (rpm)	400.0					
	π (bar)	0.042					
	R (%)	40.0	9.8	40.0	9.8	40.0	9.8
Boundary flux data	Boundary flux type	TH	TH	TH	TH	TH	TH
	α (L h^{-2} m^{-2} bar^{-1})	2.9	5.5	4.5	2.8	2.6	1.9
	Δw% (%)	N/A	N/A	N/A	N/A	N/A	N/A
	J_b (L h^{-1} m^{-2})	19.6	40.1	29.0	13.1	20.0	57.8
	TMP$_b$ (bar)	5.0	5.0	5.0	5.0	5.0	5.0

TABLE 7.45 Landfill Leachate MBR Wastewater

Reference ID		[93]			
Feedstock	Key parameter	COD			
	Value in feed stream	2070 mg L⁻¹			
	Pretreatments	Membrane bioreactor (MBR)			
Membrane properties	Membrane type	NF	NF	MF + PAC (0.1 g/L)	
	Membrane model	Flat-sheet hydrophilic PES		Hydrophilic cellulose nitrate	
	Membrane ID	FM NP010	FM NP030	N/A	N/A
	Membrane supplier	Microdyn Nadir Gmbh		Schleicher & Schuell	
	Pore size	1 kDa	0.4 kDa	200 nm	450 nm
	w (L h⁻¹ m⁻² bar⁻¹)	200.0	40.0	N/A	N/A
Process properties	T (°C)	25.0			
	v_F (m s⁻¹)	1.1			
	π (bar)	0.0426			
	R (%)	42.1	51.7	66.2	68.6
Boundary flux data	Boundary flux type	TH	TH	TH	TH
	α (L h⁻² m⁻² bar⁻¹)	1.60	0.55	10.08	15.84
	Δw% (%)	N/A	N/A	N/A	N/A
	J_b (L h⁻¹ m⁻²)	106.9	32.4	30.4	31.1
	TMP_b (bar)	20.0	20.0	17.5	7.5

TABLE 7.46 Latex Suspension

Reference ID		[5]	
Feedstock	Key parameter	Latex concentration (particle size 123 nm)	Latex concentration (particle size 190 nm)
	Value in feed stream	0.71 mg L^{-1}	4.90 mg L^{-1}
	Pretreatments	None	
Membrane properties	Membrane type	UF	MF
	Membrane model	Tubular ceramic	Tubular ceramic
	Membrane ID	Carbosep	Kerasep
	Membrane supplier	Orelis	Orelis
	Pore size	10 kDa	100 nm
	w (L h^{-1} m^{-2} bar^{-1})	72.0	49.9
Process properties	T (°C)	25.0	50.0
	v_f (m s^{-1})	0.6	0.5
	π (bar)	0.0	
	R (%)	99.9	
Boundary flux data	Boundary flux type	C	C
	α (L h^{-2} m^{-2} bar^{-1})	–	–
	Δw% (%)	N/A	N/A
	J_b (L h^{-1} m^{-2})	25.0	35.0
	TMP$_b$ (bar)	1.0	1.0

TABLE 7.47 Lead Solution

Reference ID		[94]
Feedstock	Key parameter	Lead concentration
	Value in feed stream	250 mg L^{-1}
	Pretreatments	None
Membrane properties	Membrane type	NF
	Membrane model	Thin-film composite, tubular aromatic polyamide skin layer on polysulfone substrate
	Membrane ID	AFC 80
	Membrane supplier	PCI Membranes
	Pore size	0.26 nm
	w (L h^{-1} m^{-2} bar^{-1})	1.45
Process properties	T (°C)	25.0
	v_F (m s^{-1})	1.25
	π (bar)	0.082
	R (%)	98.0
Boundary flux data	Boundary flux type	C
	α (L h^{-2} m^{-2} bar^{-1})	–
	$\Delta w\%$ (%)	–
	J_b (L h^{-1} m^{-2})	36.5
	TMP_b (bar)	3.0

TABLE 7.48 Malt Extract Wastewater

Reference ID		[95]
Feedstock	Key parameter	TSS
	Value in feed stream	132 mg L^{-1}
	Pretreatments	None
Membrane properties	Membrane type	MF
	Membrane model	Tubular Al–SiO
	Membrane ID	AB10VSX
	Membrane supplier	Atlas
	Pore size	450 nm
	w (L h^{-1} m^{-2} bar^{-1})	N/A
Process properties	T (°C)	N/A
	v_F (m s^{-1})	0.8
	π (bar)	0.0
	R (%)	99.7
Boundary flux data	Boundary flux type	C
	α (L h^{-2} m^{-2} bar^{-1})	—
	Δw% (%)	—
	J_b (L h^{-1} m^{-2})	80.0
	TMP$_b$ (bar)	1.8

TABLE 7.49 Microalgae Suspension of *Chlorella*

Reference ID		[96]
Feedstock	Key parameter	Microalgae concentration (2–5 µm)
	Value in feed stream	130.14 mg L^{-1} sodium acetate trihydrate
	Pretreatments	None
Membrane properties	Membrane type	UF
	Membrane model	Flat-sheet cellulose acetate
	Membrane ID	N/A
	Membrane supplier	GE Water and Process Technologies
	Pore size	20 kDa
	w (L h^{-1} m^{-2} bar^{-1})	N/A
Process properties	T (°C)	26.0
	v$_F$ (m s^{-1})	N/A
	π (bar)	0.0
	R (%)	N/A
Boundary flux data	Boundary flux type	TH
	α (L h^{-2} m^{-2} bar^{-1})	208.0
	Δw% (%)	N/A
	J$_b$ (L h^{-1} m^{-2})	31.2
	TMP$_b$ (bar)	1.5

TABLE 7.50 Microalgae Suspension of *Chlamydomonas*

Reference ID		[97]
Feedstock	Key parameter	Algae concentration
	Value in feed stream	0.6 g L^{-1}
	Pretreatments	None
Membrane properties	Membrane type	MF
	Membrane model	Flat sheet
	Membrane ID	GSWP
	Membrane supplier	Millipore
	Pore size	550 nm
	w (L h^{-1} m^{-2} bar^{-1})	N/A
Process properties	T (°C)	N/A
	v$_F$ (m s^{-1})	N/A
	π (bar)	0.0
	R (%)	99.9
Boundary flux data	Boundary flux type	C
	α (L h^{-2} m^{-2} bar^{-1})	–
	Δw% (%)	–
	J$_b$ (L h^{-1} m^{-2})	14.0
	TMP$_b$ (bar)	0.05

TABLE 7.51 Microalgae Suspension of *Cylindrotheca fusiformis*

Reference ID		[97]	
Feedstock	Key parameter	Algae concentration	
	Value in feed stream	2.4×10^6 cells L^{-1}	0.5×10^6 cells L^{-1}
	Pretreatments	None	
Membrane properties	Membrane type	UF	
	Membrane model	Flat-sheet polyacrilonitrile (PAN)	
	Membrane ID	IRIS 3038	
	Membrane supplier	Orelis	
	Pore size	40 kDa	
	w (L h^{-1} m^{-2} bar^{-1})	N/A	
Process properties	T (°C)	25.0	
	v_F (m s^{-1})	1.0	
	π (bar)	0.0	
	R (%)	N/A	N/A
Boundary flux data	Boundary flux type	TH	TH
	α (L h^{-2} m^{-2} bar^{-1})	71.7	170.1
	Δw% (%)	N/A	N/A
	J_b (L h^{-1} m^{-2})	41.1	63.3
	TMP$_b$ (bar)	1.0	1.0

TABLE 7.52 Modified Skim Milk Solution

Reference ID		[98]
Feedstock	Key parameter	Proteins concentration
	Value in feed stream	31.5 mg L^{-1}
	Pretreatments	pH adjustment to 11.5
Membrane properties	Membrane type	UF
	Membrane model	Spiral wound PES
	Membrane ID	HFK131
	Membrane supplier	Koch Inc.
	Pore size	N/A
	w (L h^{-1} m^{-2} bar^{-1})	10.0
Process properties	T (°C)	25.0
	v$_F$ (m s^{-1})	0.3
	π (bar)	0.0
	R (%)	93.0
Boundary flux data	Boundary flux type	C
	α (L h^{-2} m^{-2} bar^{-1})	–
	Δw% (%)	–
	J$_b$ (L h^{-1} m^{-2})	25.0
	TMP$_b$ (bar)	2.2

TABLE 7.53 Myoglobin Solution

Reference ID	Key parameter	[63]		
Feedstock		Myoglobin concentration		
	Value in feed stream	100 mg L^{-1}	200 mg L^{-1}	300 mg L^{-1}
	Pretreatments	pH adjustment to 8.0		
Membrane properties	Membrane type	UF		
	Membrane model	Flat sheet		
	Membrane ID	C30G		
	Membrane supplier	Hoechst Company (Germany)		
	Pore size	30 kDa		
	w (L h^{-1} m^{-2} bar^{-1})	40.0		
Process properties	T (°C)	25.0		
	v_F (m s^{-1})	0.19		
	π (bar)	–		
	R (%)	30.0		
Boundary flux data	Boundary flux type	C	C	C
	α (L h^{-2} m^{-2} bar^{-1})	–	–	–
	Δw% (%)	–	–	–
	J_b (L h^{-1} m^{-2})	105.0	85.0	65.0
	TMP$_b$ (bar)	0.17	0.17	0.17

TABLE 7.54 Natural Rubber Skim Latex Suspension

Reference ID		[99]
Feedstock	Key parameter	Dry rubber concentration
	Value in feed stream	4%
	Pretreatments	Sonication
Membrane properties	Membrane type	UF
	Membrane model	Tubular polyvinylidene fluoride (PVDF)
	Membrane ID	MICRO240
	Membrane supplier	PCI
	Pore size	500 nm
	w (L h^{-1} m^{-2} bar^{-1})	55.0
Process properties	T (°C)	29.0
	F_F (L h^{-1})	748.8
	π (bar)	0.0
	R (%)	N/A
Boundary flux data	Boundary flux type	C
	α (L h^{-2} m^{-2} bar^{-1})	–
	Δw% (%)	–
	J_b (L h^{-1} m^{-2})	40.0
	TMP$_b$ (bar)	2.0

TABLE 7.55 Natural Organic Matter Solution (NOM)

Reference ID		[100]			
Feedstock	Key parameter	TOC			
	Value in feed stream	$10\,mg\,L^{-1}$ NOM + $9.3\,meq\,L^{-1}$ NaCl	$10\,mg\,L^{-1}$ NOM + $6.2\,meq\,L^{-1}$ $CaCl_2$	$10\,mg\,L^{-1}$ NOM + $4.7\,meq\,L^{-1}$ $CaSO_4$	$10\,mg\,L^{-1}$ NOM + $3.7\,meq\,L^{-1}$ $Ca_3(PO_4)_2$
	Pretreatments	None			
Membrane properties	Membrane type	NF			
	Membrane model	Flat-sheet PA/PS TFC			
	Membrane ID	HL 2540F1072			
	Membrane supplier	GE Osmonics, Inc.			
	Pore size	300 Da			
	w ($L\,h^{-1}\,m^{-2}\,bar^{-1}$)	13.7	24.2	54.2	72.7
Process properties	T (°C)	25.0			
	v_F ($m\,s^{-1}$)	0.1			
	π (bar)	0.056	0.05	0.046	0.045
	R (%)	94.9	97.0	93.3	93.0
Boundary flux data	Boundary flux type	TH	TH	TH	TH
	α ($L\,h^{-2}\,m^{-2}\,bar^{-1}$)	1.3	3.0	2.4	2.4
	$\Delta w\%$ (%)	–	–	–	–
	J_b ($L\,h^{-1}\,m^{-2}$)	33.3	29.7	26.9	25.6
	TMP_b (bar)	3.0	3.0	3.0	3.0

TABLE 7.56 Nickel Solution

	Reference ID	[59]		
Feedstock	Key parameter	Nickel concentration		
	Value in feed stream	10 ppm		
	Pretreatments	SDS addition		
Membrane properties	Membrane type	Micellar-enhanced UF		
	Membrane model	Polysulfone	PES	Polysulfone
	Membrane ID	–	NP010	UFX5
	Membrane supplier	Lab Made	Microdyn®	Alfa Laval
	Pore size	5 kDa	1 kDa	5 kDa
	w (L h^{-1} m^{-2} bar^{-1})	65.0	16.0	60.0
Process properties	T (°C)	25.0		
	v_F (m s^{-1})	N/A	N/A	N/A
	π (bar)	0.0	0.0	0.0
	R (%)	92	98.0	94
Boundary flux data	Boundary flux type	C	C	C
	α (L h^{-2} m^{-2} bar^{-1})	–	–	–
	$\Delta w\%$ (%)	–	–	–
	J_b (L h^{-1} m^{-2})	44.6	8.3	29.3
	TMP_b (bar)	2.5	10.0	2.5

TABLE 7.57 Nitrate Solution

Reference ID		[101]		
Feedstock	Key parameter	Nitrate concentration		
	Value in feed stream	100–300 mg L^{-1}		
	Pretreatments	None		
Membrane properties	Membrane type	NF		
	Membrane model	Flat-sheet thin-film composite polyamide/polysulfone	Flat-sheet TFC	Flat-sheet thin-film composite polyamide
	Membrane ID	NF90	NF270	ESNA1-LF
	Membrane supplier	Dow FilmTec		Hydranautics
	Pore size	N/A	N/A	N/A
	w (L h^{-1} m^{-2} bar^{-1})	3.78	9.29	7.31
Process properties	T (°C)	25.0		
	v$_F$ (m s^{-1})	2.0		
	π (bar)	0.65		
	R (%)	90.0	60.0	95.0
Boundary flux data	Boundary flux type	C	C	C
	α (L h^{-2} m^{-2} bar^{-1})	–	–	–
	Δw% (%)	–	–	–
	J$_b$ (L h^{-1} m^{-2})	18.07	33.05	42.1
	TMP$_b$ (bar)	16.0	16.0	16.0

TABLE 7.58 Oil-in-Water Emulsion

Reference ID		[102]				
Feedstock	Key parameter	Oil concentration				
	Value in feed stream	10%	20%	30%	40%	0.9 g L^{-1}
	Pretreatments	HCl addition (small amount)				None
Membrane properties	Membrane type	UF				MF
	Membrane model	Multichannel ceramic				Spherical-fly ash-based ceramic
	Membrane ID	–				–
	Membrane supplier	Tami				Lab Made
	Pore size	300 kDa				770 nm
	w (L h^{-1} m^{-2} bar^{-1})	60.0	80.0	60.0	80.0	15,600.0
Process properties	T (°C)	30.0	70.0	30.0	70.0	20.0
	v$_F$ (m s^{-1})	4.0				
	π (bar)	0.0	0.0	0.0	0.0	0.0
	R (%)	99.9	99.6	99.9	99.7	99.5
Boundary flux data	Boundary flux type	C	C	C	C	C
	α (L h^{-2} m^{-2} bar^{-1})	–	–	–	–	–
	Δw% (%)	–	–	–	–	–
	J$_b$ (L h^{-1} m^{-2})	30.0	35.0	21.0	25.0	130.0
	TMP$_b$ (bar)	0.50	0.43	1.10	0.71	2.50

TABLE 7.59 Oily Industrial Wastewater from Animal Industry

Reference ID		[103]	
Feedstock	Key parameter	COD	
	Value in feed stream	29 g L^{-1}	
	Pretreatments	None	
Membrane properties	Membrane type	UF	UF
	Membrane model	Ultrahydrofilic polymeric (polyacrylonitrile)	Cellulose acetate (hydrolyzed)
	Membrane ID	M-series	N/A
	Membrane supplier	GE Osmonics Septa	Hydration Technology Innovations
	Pore size	100 kDa	100 kDa
	w (L h^{-1} m^{-2} bar^{-1})	1.9	0.9
Process properties	T (°C)	25.0	
	v$_F$ (m s^{-1})	N/A	
	π (bar)	0.0	0.0
	R (%)	70.0	84.0
Boundary flux data	Boundary flux type	TH	TH
	α (L h^{-2} m^{-2} bar^{-1})	4.90	0.29
	Δw% (%)	N/A	N/A
	J$_b$ (L h^{-1} m^{-2})	75.6	14.4
	TMP$_b$ (bar)	0.7	2.8

TABLE 7.60 Oily Wastewater

Reference ID		[104]
Feedstock	Key parameter	Oil-in-water emulsion concentration
	Value in feed stream	200 mg L^{-1}
	Pretreatments	None
Membrane properties	Membrane type	UF
	Membrane model	Composite ceramic–polymeric
	Membrane ID	N/A
	Membrane supplier	N/A
	Pore size	28 nm
	w (L h^{-1} m^{-2} bar^{-1})	1.59
Process properties	T (°C)	25.0
	v$_F$ (m s^{-1})	N/A
	π (bar)	0.0
	R (%)	20.2
Boundary flux data	Boundary flux type	TH
	α (L h^{-2} m^{-2} bar^{-1})	10.53
	Δw% (%)	N/A
	J$_b$ (L h^{-1} m^{-2})	9.4
	TMP$_b$ (bar)	1.36

TABLE 7.61 Olive Mill Wastewater from 2 Phase Processes

Reference ID		[105]			
Feedstock	Key parameter	COD			
	Value in feed stream	14.5 g L⁻¹	11.1 g L⁻¹	9.0 g L⁻¹	6.0 g L⁻¹
	Pretreatments	HNO₃ flocculation	HNO₃ flocculation + photocatalysis	HNO₃ flocculation + UF	HNO₃ flocculation + photocatalysis + UF
Membrane properties	Membrane type	UF		NF	
	Membrane model	spiral wounded TF		spiral wounded TF	
	Membrane ID	GM2540		DK2540	
	Membrane supplier	GE Water			
	Pore size	2 nm		0.5 nm	
	w [L h⁻¹ m⁻² bar⁻¹]	5.2		2.5	
Process properties	T [°C]	20.0			
	v_F [L h⁻¹]	550			
	π [bar]	0.0			
	R [%]	39.9	48.5	69.4	76.6
Boundary flux data	Boundary flux type	TH	TH	TH	TH
	α [L h⁻² m⁻² bar⁻¹]	0.016	0.011	0.007	0.005
	Δw% [%]	N/A	N/A	N/A	N/A
	J_b [L h⁻¹ m⁻²]	8.2	10.0	11.8	14.3
	TMP_b [bar]	9.0	10.0	8.0	9.0

TABLE 7.62 Olive Mill Wastewater from 3 Phase Processes

Reference ID		[106]			
Feedstock	Key parameter	COD			
	Value in feed stream	19.1 g L^{-1}	15.2 g L^{-1}	14.0 g L^{-1}	12.6 g L^{-1}
	Pretreatments	HNO_3 flocculation	HNO_3 flocculation + photocatalysis	HNO_3 flocculation + UF	HNO_3 flocculation + photocatalysis + UF
Membrane properties	Membrane type	UF		NF	
	Membrane model	spiral wounded TF		spiral wounded TF	
	Membrane ID	GM2540		DK2540	
	Membrane supplier	GE Water			
	Pore size	2 nm		0.5 nm	
	w [L h^{-1} m^{-2} bar^{-1}]	5.2		2.5	
Process properties	T [°C]	20.0			
	v_F [L h^{-1}]	600			
	π [bar]	0.0		0.9	
	R [%]	25.0	23.0	74.0	76.0
Boundary flux data	Boundary flux type	TH	TH	TH	TH
	α [L h^{-2} m^{-2} bar^{-1}]	0.0239	0.0553	0.0084	0.0191
	Δw% [%]	0.019	0.016	0.012	0.009
	J_b [L h^{-1} m^{-2}]	3.1	7.6	12.2	14.9
	TMP_b [bar]	4.0	6.0	4.0	8.0

TABLE 7.63 *Pseudomonas putida* Cells Suspension

Reference ID		[107]	
Feedstock	Key parameter	*P. putida* cells concentration	
	Value in feed stream	1.6×10^4 cells mL^{-1}	
	Pretreatments	None	
Membrane properties	Membrane type	UF	
	Membrane model	Flat-sheet polysulfone treated by TiO_2 and UV irradiation	Flat-sheet polysulfone
	Membrane ID	PS100	
	Membrane supplier	Microdyn Nadir GmbH	
	Pore size	0.1 kDa	
	w (L h^{-1} m^{-2} bar^{-1})	N/A	
Process properties	T (°C)	25.0	
	v_F (m s^{-1})	N/A	
	π (bar)	N/A	
	R (%)	N/A	
Boundary flux data	Boundary flux type	TH	TH
	α (L h^{-2} m^{-2} bar^{-1})	1.00	0.35
	Δw% (%)	N/A	N/A
	J_b (L h^{-1} m^{-2})	564.28	255.1
	TMP$_b$ (bar)	3.0	3.0

TABLE 7.64 Paper Industry Wastewater

Reference ID		[4]
Feedstock	Key parameter	TOC
	Value in feed stream	430 mg L^{-1}
	Pretreatments	None
Membrane properties	Membrane type	NF
	Membrane model	Spiral wound PA membrane TF
	Membrane ID	Desal5
	Membrane supplier	GE Osmonics
	Pore size	2 nm
	w (L h^{-1} m^{-2} bar^{-1})	2.25
Process properties	T (°C)	40.0
	v$_F$ (m s^{-1})	2.0
	π (bar)	0.17
	R (%)	N/A
Boundary flux data	Boundary flux type	C
	α (L h^{-2} m^{-2} bar^{-1})	–
	Δw% (%)	–
	J$_b$ (L h^{-1} m^{-2})	35.0
	TMP$_b$ (bar)	6.0

TABLE 7.65 Polyethylene Glycol Solution (PEG)

Reference ID		[108]		
Feedstock	Key parameter	PEG (20 kDa) concentration		
	Value in feed stream	$1\ g\ L^{-1}$	$0.05\ g\ L^{-1}$	$0.1\ g\ L^{-1}$
	Pretreatments	None		
Membrane properties	Membrane type	MF		
	Membrane model	Disc ceramic	Flat-sheet polycarbonate (PC)	
	Membrane ID			
	Membrane supplier	Sterlitech	Millipore	
	Pore size	200 nm	200 nm	200 nm
	w ($L\ h^{-1}\ m^{-2}\ bar^{-1}$)	N/A	N/A	N/A
Process properties	T (°C)	23.0		
	v_F ($m\ s^{-1}$)	N/A		
	π (bar)	0.0	0.0	0.0
	R (%)	100.0	100.0	100.0
Boundary flux data	Boundary flux type	C	TH	TH
	α ($L\ h^{-2}\ m^{-2}\ bar^{-1}$)	–	202.3	63.2
	$\Delta w\%$ (%)	–	N/A	N/A
	J_b ($L\ h^{-1}\ m^{-2}$)	20.88	32.88	21.92
	TMP_b (bar)	0.13	0.13	0.13

TABLE 7.66 Petrochemical Wastewaters

Reference ID		[109]	
Feedstock	Key parameter	Coke particles (5–30 μm), COD	
	Value in feed stream	2210–1.9 × 10^5 particles mm^{-3}, 2.5 × 10^5 mg L^{-1}	
	Pretreatments	None	Guard filter
Membrane properties	Membrane type	MF	
	Membrane model	Tubular γ-Al$_2$O$_3$ ceramic (7 channels, 1.5–2 mm thickness, 1 mL, 2.5 cm outer D)	
	Membrane ID	N/A	
	Membrane supplier	Iran Membrane Co.	
	Pore size	200 nm	
	w (L h^{-1} m^{-2} bar^{-1})	390.0	
Process properties	T (°C)	60.0	
	v$_F$ (m s^{-1})	2.0	
	π (bar)	0.0	0.0
	R (%)	100.0	100.0
Boundary flux data	Boundary flux type	C	C
	α (L h^{-2} m^{-2} bar^{-1})	–	–
	Δw% (%)	–	–
	J$_b$ (L h^{-1} m^{-2})	666.0	
	TMP$_b$ (bar)	15.0	

TABLE 7.67 Pharmaceutical Industry Wastewater

Reference ID		[110]			
Feedstock	Key parameter	TOC			
	Value in feed stream	2.8 g L⁻¹			
	Pretreatments	None		Anthracite–sand biofiltration	
Membrane properties	Membrane type	NF			
	Membrane model	Spiral wound polyamide-urea TFC	Spiral wound aromatic polyamide TFC	Spiral wound polyamide-urea TFC	Spiral wound aromatic polyamide TFC
	Membrane ID	Sepa CF XN45	Sepa CF TS80	Sepa CF XN45	Sepa CF TS80
	Membrane supplier	TriSep GE Osmonics			
	Pore size	200 Da	100 Da	200 Da	100 Da
	w (L h⁻¹ m⁻² bar⁻¹)	10.7	6.7	10.7	6.7
Process properties	T (°C)	20.0			
	v_F (m s⁻¹)	0.4			
	π (bar)	0.12		0.10	
	R (%)	71.0	88.0	73.0	96.0
Boundary flux data	Boundary flux type	TH	TH	TH	TH
	α (L h⁻² m⁻² bar⁻¹)	0.067	0.32	0.052	0.14
	Δw% (%)	N/A	N/A	N/A	N/A
	J_b (L h⁻¹ m⁻²)	90.0	70.0	50.0	15.0
	TMP_b (bar)	4.8	4.8	4.8	4.8

TABLE 7.68 Phenol Solution

Reference ID		[111]
Feedstock	Key parameter	TOC
	Value in feed stream	80 mg L^{-1}
	Pretreatments	Wet peroxide oxidation
Membrane properties	Membrane type	NF
	Membrane model	Spiral wound polymeric TFC PA/PS
	Membrane ID	Filmtec NF90
	Membrane supplier	Dow Water
	Pore size	200 Da
	w (L h^{-1} m^{-2} bar^{-1})	5.6
Process properties	T (°C)	30.0
	v$_F$ (m s^{-1})	0.083
	π (bar)	0.042
	R (%)	30.0
Boundary flux data	Boundary flux type	TH
	α (L h^{-2} m^{-2} bar^{-1})	0.11
	Δw% (%)	N/A
	J$_b$ (L h^{-1} m^{-2})	31.4
	TMP$_b$ (bar)	6.0

TABLE 7.69 Plasmid DNA Solution

Reference ID		[112]
Feedstock	Key parameter	Plasmid DNA concentration
	Value in feed stream	0.33 µg L^{-1}
	Pretreatments	None
Membrane properties	Membrane type	UF
	Membrane model	Flat-sheet PES
	Membrane ID	N/A
	Membrane supplier	Sterlitech Corporation
	Pore size	20 kDa
	w (L h^{-1} m^{-2} bar^{-1})	40.32
Process properties	T (°C)	20.0
	v$_F$ (rpm)	400
	π (bar)	0.0
	R (%)	99.92
Boundary flux data	Boundary flux type	TH
	α (L h^{-2} m^{-2} bar^{-1})	0.4
	Δw% (%)	N/A
	J$_b$ (L h^{-1} m^{-2})	32.25
	TMP$_b$ (bar)	1.0

TABLE 7.70 Polymer Flooding in Oil Solution Wastewater

Reference ID		[113]	
Feedstock	Key parameter	Nonionic PAM (10^4 kDa)	
	Value in feed stream	5 mg L^{-1}	
	Pretreatments	None	
Membrane properties	Membrane type	UF	MF
	Membrane model	Tubular ceramic 19 channels ZrO_2/α-Al_2O_3	
	Membrane ID	N/A	
	Membrane supplier	Jiangsu Jiuwu Hitech Co.	
	Pore size	200 nm	500 nm
	w (L h^{-1} m^{-2} bar^{-1})	1074.2	1330.4
Process properties	T (°C)	25.0	
	v_F (m s^{-1})	2.0	
	π (bar)	0.0	0.0
	R (%)	N/A	N/A
Boundary flux data	Boundary flux type	C	C
	α (L h^{-1} m^{-2} bar^{-1})	–	–
	Δw% (%)	–	–
	J_b (L h^{-1} m^{-2})	177.8	253.5
	TMP_b (bar)	1.0	1.0

TABLE 7.71 Polymeric Suspension

Reference ID		[113]
Feedstock	Key parameter	Particle (1.3 μm) concentration
	Value in feed stream	1 g L^{-1}
	Pretreatments	None
Membrane properties	Membrane type	MF
	Membrane model	Spherical fly ash-based ceramic
	Membrane ID	N/A
	Membrane supplier	Lab Made
	Pore size	770 nm
	w (L h^{-1} m^{-2} bar^{-1})	1560.0
Process properties	T (°C)	20.0
	v$_F$ (m s^{-1})	4.0
	π (bar)	0.0
	R (%)	99.9
Boundary flux data	Boundary flux type	TH
	α (L h^{-2} m^{-2} bar^{-1})	16.7
	Δw% (%)	N/A
	J$_b$ (L h^{-1} m^{-2})	420.0
	TMP$_b$ (bar)	3.0

TABLE 7.72 Polymethyl Methacrylate Solution

Reference ID		[115]			
Feedstock	Key parameter	PMMA particles (0.15 mm) concentration			
	Value in feed stream	0.5 g L^{-1}		2 g L^{-1}	
	Pretreatments	None			
Membrane properties	Membrane type	MF			
	Membrane model	Hydrophilic polycarbonate flat sheet			
	Membrane ID	Isopore1			
	Membrane supplier	Millipore Co.			
	Pore size	200 nm	400 nm	200 nm	400 nm
	w (L h^{-1} m^{-2} bar^{-1})	246.9			
Process properties	T (°C)	20.0			
	v$_F$ (m s^{-1})	N/A			
	π (bar)	0.0			
	R (%)	N/A			
Boundary flux data	Boundary flux type	TH	TH	TH	TH
	α (L h^{-2} m^{-2} bar^{-1})	35.9	23.6	3.5	7.1
	Δw% (%)	N/A	N/A	N/A	N/A
	J$_b$ (L h^{-1} m^{-2})	136.45	113.71	61.73	48.73
	TMP$_b$ (bar)	5.0	5.0	5.0	5.0

TABLE 7.73 Polystyrene Latex Suspension

Reference ID		[116]	
Feedstock	Key parameter	Polystyrene latex concentration (particle size 0.1 μm)	
	Value in feed stream	10 mg L^{-1}	
	Pretreatments	None	
Membrane properties	Membrane type	MF	
	Membrane model	Polyvinylidene fluoride hydrophobic	
	Membrane ID	Modified PVDF	Unmodified PVDF
	Membrane supplier	Millipore, USA	
	Pore size	325 nm	220 nm
	w (L h^{-1} m^{-2} bar^{-1})	N/A	N/A
Process properties	T (°C)	25.0	
	v$_F$ (m s^{-1})	N/A	
	π (bar)	N/A	N/A
	R (%)	N/A	N/A
Boundary flux data	Boundary flux type	TH	TH
	α (L h^{-2} m^{-2} bar^{-1})	0.021	0.033
	Δw% (%)	N/A	N/A
	J$_b$ (L h^{-1} m^{-2})	357.0	73.5
	TMP$_b$ (bar)	1.4	1.4

TABLE 7.74 *Pseudomonas aeruginosa* Suspension

Reference ID		[117]
Feedstock	Key parameter	*Pseudomonas* concentration
	Value in feed stream	10^8 cfu L^{-1}
	Pretreatments	30 µM D-tyrosine
Membrane properties	Membrane type	NF
	Membrane model	Flat sheet
	Membrane ID	NF270
	Membrane supplier	Dow Filmtec
	Pore size	N/A
	w ($L\ h^{-1}\ m^{-2}\ bar^{-1}$)	N/A
Process properties	T (°C)	20.0
	v_F ($m\ s^{-1}$)	N/A
	π (bar)	0.35
	R (%)	99.9
Boundary flux data	Boundary flux type	TH
	α ($L\ h^{-2}\ m^{-2}\ bar^{-1}$)	0.0007
	Δw% (%)	N/A
	J_b ($L\ h^{-1}\ m^{-2}$)	0.12
	TMP_b (bar)	6.0

TABLE 7.75 Raw Rice Wine

Reference ID	Key parameter	[118]		
Feedstock	Key parameter	Crude protein concentration		
	Value in feed stream	10.8 g L^{-1}		
	Pretreatments	None		
Membrane properties	Membrane type	MF		
	Membrane model	Ceramic tubular α-Al$_2$O$_3$		Ceramic tubular ZrO$_2$
	Membrane ID	N/A	N/A	N/A
	Membrane supplier	Nanjing Jiusi High-Tech Co., Ltd, Jiangsu, PR China		
	Pore size	200 nm	500 nm	200 nm
	w (L h^{-1} m^{-2} bar^{-1})	177.4		N/A
Process properties	T (°C)	15.0		
	v_F (m s^{-1})	1.98		
	π (bar)	0.0	0.0	0.0
	R (%)	74.2	60	76.3
Boundary flux data	Boundary flux type	TH	TH	TH
	α (L h^{-2} m^{-2} bar^{-1})	6.7	4.3	4.96
	Δw% (%)	N/A	N/A	N/A
	J_b (L h^{-1} m^{-2})	54.3	18.0	33.7
	TMP$_b$ (bar)	1.0	1.0	1.0

TABLE 7.76 Raw River Water

Reference ID		[119]			
Feedstock	Key parameter	COD			
	Value in feed stream	9.7 mg L^{-1}		10.6 mg L^{-1}	
	Pretreatments	pH adjustment to 2.0 with H$_2$SO$_4$ and 0.45 μm prefilter			
Membrane properties	Membrane type	UF			
	Membrane model	Tubular ceramic PVDF	Tubular ceramic PVDF modified by nanoalumina	Tubular ceramic PVDF	Tubular ceramic modified PVDF by nanoalumina
	Membrane ID	N/A	N/A	N/A	N/A
	Membrane supplier	N/A	N/A	N/A	N/A
	Pore size	35 kDa	35 kDa	35 kDa	35 kDa
	w (L h^{-1} m^{-2} bar^{-1})	30.0	310.0	38.0	310.0
Process properties	T (°C)	25.0			
	v_F (m s^{-1})	0.3			
	π (bar)	0.0	0.0	0.0	0.0
	R (%)	N/A	N/A	N/A	N/A
Boundary flux data	Boundary flux type	C	C	C	C
	α (L h^{-2} m^{-2} bar^{-1})	–	–	–	–
	Δw% (%)	–	–	–	–
	J_b (L h^{-1} m^{-2})	6.0	173.3	6.0	155.0
	TMP$_b$ (bar)	1.0	1.0	1.0	1.0

TABLE 7.77 Red Wine

Reference ID		[120]	
Feedstock	Key parameter	Tannins concentration	
	Value in feed stream	1.25 g L^{-1}	2.5 g L^{-1}
	Pretreatments	None	
Membrane properties	Membrane type	MF	
	Membrane model	Multichannel ceramic module	
	Membrane ID	BK compact	
	Membrane supplier	Novasep	
	Pore size	200 nm	
	w (L h^{-1} m^{-2} bar^{-1})	900.0	
Process properties	T (°C)	20.0	
	v$_F$ (m s^{-1})	2.0	
	π (bar)	0.0	0.0
	R (%)	N/A	N/A
Boundary flux data	Boundary flux type	C	C
	α (L h^{-2} m^{-2} bar^{-1})	—	—
	Δw% (%)	—	—
	J$_b$ (L h^{-1} m^{-2})	90.2	5.7
	TMP$_b$ (bar)	0.78	0.86

TABLE 7.78 Rough Nonalcoholic Beer

Reference ID		[95]
Feedstock	Key parameter	TSS
	Value in feed stream	26 g L^{-1}
	Pretreatments	None
Membrane properties	Membrane type	MF
	Membrane model	Tubular ceramic Al–SiO
	Membrane ID	AB10VSX
	Membrane supplier	Atlas
	Pore size	450 nm
	w (L h^{-1} m^{-2} bar^{-1})	N/A
Process properties	T (°C)	7.0
	v_F (m s^{-1})	0.8
	π (bar)	0.0
	R (%)	99.5
Boundary flux data	Boundary flux type	C
	α (L h^{-2} m^{-2} bar^{-1})	—
	Δw% (%)	—
	J_b (L h^{-1} m^{-2})	144.0
	TMP$_b$ (bar)	1.1

TABLE 7.79 Safranin T Solution

Reference ID	Key parameter	[121]				
Feedstock		Safranin T concentration (MW 350.85)				
		1 mM		5 mM		
	Value in feed stream	None	Anionic surfactant SDS addition	None	Anionic surfactant SDS addition	
	Pretreatments	None	Anionic surfactant SDS addition	None	Anionic surfactant SDS addition	
Membrane properties	Membrane type	Micellar-enhanced UF	UF	Micellar-enhanced UF	UF	
	Membrane model	N/A	N/A	N/A	N/A	
	Membrane ID	Flat-sheet polymeric regenerated cellulose				
	Membrane supplier	Millipore				
	Pore size	10 kDa				
	w (L h^{-1} m^{-2} bar^{-1})	122.3	108.7	119	114.2	
Process properties	T (°C)	22.0				
	v_F (m s^{-1})	N/A				
	π (bar)	0.0	0.0	0.0	0.0	
	R (%)	99.0	99.0	99.0	99.0	
Boundary flux data	Boundary flux type	TH	TH	TH	TH	
	α (L h^{-2} m^{-2} bar^{-1})	2.6	0.7	0.2	0.4	
	$\Delta w\%$ (%)	N/A	N/A	N/A	N/A	
	J_b (L h^{-1} m^{-2})	141.1	99.9	133.4	85.5	
	TMP_b (bar)	1.4	1.4	1.4	1.4	

TABLE 7.80 Seawater

Reference ID		[122]
Feedstock	Key parameter	COD
	Value in feed stream	2.6 mg L^{-1}
	Pretreatments	0.5 mg L^{-1} FeCl$_3$; 1.5 g L^{-1} PAC
Membrane properties	Membrane type	MF
	Membrane model	HF PVDF
	Membrane ID	S
	Membrane supplier	CleanFil
	Pore size	100 nm
	w (L h^{-1} m^{-2} bar^{-1})	N/A
Process properties	T (°C)	20.0
	v$_F$ (m s^{-1})	N/A
	π (bar)	0.0
	R (%)	72.0
Boundary flux data	Boundary flux type	C
	α (L h^{-2} m^{-2} bar^{-1})	–
	Δw% (%)	–
	J$_b$ (L h^{-1} m^{-2})	70.0
	TMP$_b$ (bar)	2.8

TABLE 7.81 Secondary Municipal Effluent

Reference ID		[123]
Feedstock	Key parameter	COD
	Value in feed stream	30.6 mg L^{-1}
	Pretreatments	None
Membrane properties	Membrane type	MF
	Membrane model	Spiral wound PVDF
	Membrane ID	N/A
	Membrane supplier	Lab Made
	Pore size	150 nm
	w (L h^{-1} m^{-2} bar^{-1})	500.0
Process properties	T (°C)	18.0
	F$_F$ (L h^{-1})	2300.0
	π (bar)	0.0
	R (%)	50.1
Boundary flux data	Boundary flux type	C
	α (L h^{-2} m^{-2} bar^{-1})	–
	Δw% (%)	–
	J$_b$ (L h^{-1} m^{-2})	52.0
	TMP$_b$ (bar)	2.3

TABLE 7.82 Sequenced Batch Reactor Effluent

Reference ID		[83]
Feedstock	Key parameter	TOC
	Value in feed stream	209.5 g L^{-1}
	Pretreatments	pH adjustment to 7–8
Membrane properties	Membrane type	UF
	Membrane model	Tubular PVDF
	Membrane ID	N/A
	Membrane supplier	Ande Membrane Separation Technology and Engineering
	Pore size	100 nm
	w (L h^{-1} m^{-2} bar^{-1})	234.8
Process properties	T (°C)	25.0
	v$_F$ (m s^{-1})	N/A
	π (bar)	0.0
	R (%)	99.9
Boundary flux data	Boundary flux type	TH
	α (L h^{-2} m^{-2} bar^{-1})	0.92
	Δw% (%)	N/A
	J$_b$ (L h^{-1} m^{-2})	84.6
	TMP$_b$ (bar)	1.0

TABLE 7.83 Shell Gas Process Wastewater

Reference ID		[124]	
Feedstock	Key parameter	TDS	
	Value in feed stream	48.0 g L^{-1}	13.9 g L^{-1}
	Pretreatments	None	
Membrane properties	Membrane type	MF	
	Membrane model	Tubular ceramic	
	Membrane ID	N/A	
	Membrane supplier	Lab Made	
	Pore size	1400 nm	200 nm
	w (L h^{-1} m^{-2} bar^{-1})	3000.0	2100.0
Process properties	T (°C)	25.0	
	v_F (m s^{-1})	–	
	π (bar)	–	–
	R (%)	71.0	100.0
Boundary flux data	Boundary flux type	C	C
	α (L h^{-2} m^{-2} bar^{-1})	–	–
	Δw% (%)	–	–
	J_b (L h^{-1} m^{-2})	19.0	60.0
	TMP$_b$ (bar)	1.0	1.0

TABLE 7.84 Sewage Water

Reference ID		[125]	
Feedstock	Key parameter	Oil concentration	
	Value in feed stream	80 g L^{-1}	
	Pretreatments	None	
Membrane properties	Membrane type	UF	
	Membrane model	Sulfated Y-doped nonstoichiometric 15 wt% zirconia to PSF	Commercial PSF
	Membrane ID	N/A	N/A
	Membrane supplier	Lab Made	Dalian Polysulfone Co., Ltd
	Pore size	N/A	88.4 nm
	w (L h^{-1} m^{-2} bar^{-1})	N/A	N/A
Process properties	T (°C)	25.0	
	v$_F$ (m s^{-1})	N/A	
	π (bar)	0.0	0.0
	R (%)	99.2	98.7
Boundary flux data	Boundary flux type	TH	TH
	α (L h^{-2} m^{-2} bar^{-1})	26.6	15.0
	Δw% (%)	N/A	N/A
	J$_b$ (L h^{-1} m^{-2})	110	60
	TMP$_b$ (bar)	1.5	1.5

TABLE 7.85 Silica Particles Suspension

Reference ID		[126]
Feedstock	Key parameter	Silica particles concentration
	Value in feed stream	1.6 g L^{-1}
	Pretreatments	None
Membrane properties	Membrane type	MF
	Membrane model	Ceramic tubular membrane
	Membrane ID	Membranlox
	Membrane supplier	Bazet
	Pore size	200 nm
	w (L h^{-1} m^{-2} bar^{-1})	1500.0
Process properties	T (°C)	N/A
	v$_F$ (m s^{-1})	5.0
	π (bar)	0.0
	R (%)	99.9
Boundary flux data	Boundary flux type	C
	α (L h^{-2} m^{-2} bar^{-1})	–
	Δw% (%)	–
	J$_b$ (L h^{-1} m^{-2})	90.0
	TMP$_b$ (bar)	0.06

TABLE 7.86 Simazine Solution

Reference ID			[127]	
Feedstock	Key parameter		Simazine (207 Da) concentration	
	Value in feed stream		100 µg L^{-1}	
	Pretreatments		None	
Membrane properties	Membrane type		NF	NF
	Membrane model		Polypiperazine amide	Polyamide
	Membrane ID		FilmTec	OPMN-K
	Membrane supplier		Dow Corporation	Vladipor Society
	Pore size		170 Da	330 Da
	w (L h^{-1} m^{-2} bar^{-1})		4.5	7.9
Process properties	T (°C)		20.0	
	v$_f$ (m s^{-1})		0.45	
	π (bar)		0.0	0.0
	R (%)		96.0	69.0
Boundary flux data	Boundary flux type		C	C
	α (L h^{-2} m^{-2} bar^{-1})		–	–
	Δw% (%)		–	–
	J$_b$ (L h^{-1} m^{-2})		67.5	100.0
	TMP$_b$ (bar)		15.0	10.0

TABLE 7.87 Sodium Alginate Solution

Reference ID		[128]			
Feedstock	Key parameter	Sodium alginate (SA) (polysaccharide; 200 nm)			
	Value in feed stream	20 mg L⁻¹			
	Pretreatments	None	Polyvinyl alcohol (PVA, 100 nm) addition (20 mg L⁻¹)	None	Polyvinyl alcohol (PVA, 100 nm) addition (20 mg L⁻¹)
Membrane properties	Membrane type	MF	MF	UF	UF
	Membrane model	Flat-sheet PES			
	Membrane ID	N/A			
	Membrane supplier	Synder			
	Pore size	200 nm	200 nm	1 kDa	1 kDa
	w (L h⁻¹ m⁻² bar⁻¹)	N/A	N/A	N/A	N/A
Process properties	T (°C)	20.0			
	v_F (m s⁻¹)	N/A			
	π (bar)	0.0	0.0	0.0	0.0
	R (%)	95.0	95.0	95.0	95.0
Boundary flux data	Boundary flux type	TH	TH	TH	TH
	α (L h⁻² m⁻² bar⁻¹)	8.44	6.44	2.4	2.0
	$\Delta w\%$ (%)	N/A	N/A	N/A	N/A
	J_b (L h⁻¹ m⁻²)	7.3	5.6	7.6	5.9
	TMP_b (bar)	2.5	2.5	2.5	2.5

TABLE 7.88 Soybean Biodiesel

Reference ID		[129]	
Feedstock	Key parameter	Glycerol concentration	
	Value in feed stream	6.2 g L^{-1}	
	Pretreatments	20% of 5% HCl water	30% of 5% HCl water
Membrane properties	Membrane type	MF	
	Membrane model	Tubular ceramic	
	Membrane ID	N/A	
	Membrane supplier	Shumacher Gmbh	
	Pore size	100 nm	
	w (L h^{-1} m^{-2} bar^{-1})	N/A	
Process properties	T (°C)	50.0	
	v$_F$ (m s^{-1})	8.0	
	π (bar)	0.0	
	R (%)	99.9	99.9
Boundary flux data	Boundary flux type	C	TH
	α (L h^{-2} m^{-2} bar^{-1})	—	7.6
	Δw% (%)	N/A	N/A
	J$_b$ (L h^{-1} m^{-2})	6.3	74.0
	TMP$_b$ (bar)	1.0	2.0

TABLE 7.89 Synthetic Oily Wastewater

Reference ID		[130]
Feedstock	Key parameter	COD
	Value in feed stream	510 mg L^{-1}
	Pretreatments	None
Membrane properties	Membrane type	MF
	Membrane model	N/A
	Membrane ID	N/A
	Membrane supplier	N/A
	Pore size	200 nm
	w (L h^{-1} m^{-2} bar^{-1})	35.0
Process properties	T (°C)	25.0
	v$_F$ (m s^{-1})	2.0
	π (bar)	0.0
	R (%)	87.45
Boundary flux data	Boundary flux type	C
	α (L h^{-2} m^{-2} bar^{-1})	—
	Δw% (%)	—
	J$_b$ (L h^{-1} m^{-2})	67.1
	TMP$_b$ (bar)	3.0

TABLE 7.90 Textile Industry Wastewater

Reference ID		[131]	
Feedstock	Key parameter	COD	
	Value in feed stream	627 mg L^{-1}	
	Pretreatments	Coag. Al$_2$(SO$_4$)$_3$16H$_2$O-flocc. polyelectrolyte magnafloc 919	MF 200 nm
Membrane properties	Membrane type	NF	
	Membrane model	Spiral wound polymeric PA/PS TFC	
	Membrane ID	DK	
	Membrane supplier	GE Osmonics	
	Pore size	0.2 nm	
	w (L h^{-1} m^{-2} bar^{-1})	4.0	
Process properties	T (°C)	40.0	
	v$_F$ (m s^{-1})	5.6	
	π (bar)	0.68	
	R (%)	51.8	50.8
Boundary flux data	Boundary flux type	TH	TH
	α (L h^{-2} m^{-2} bar^{-1})	1.18	9.5
	Δw% (%)	N/A	N/A
	J$_b$ (L h^{-1} m^{-2})	13.5	34.5
	TMP$_b$ (bar)	10.0	2.0

TABLE 7.91 Titanium Dioxide Suspension

Reference ID			[132]
Feedstock		Key parameter	TiO$_2$ suspension concentration (mean size 300 nm)
		Value in feed stream	0.01 g L^{-1}
		Pretreatments	None
Membrane properties		Membrane type	MF
		Membrane model	Tubular ceramic
		Membrane ID	N/A
		Membrane supplier	Fairey Ind.
		Pore size	N/A
		w (L h^{-1} m^{-2} bar^{-1})	300.0
Process properties		T (°C)	25.0
		v$_F$ (m s^{-1})	2.2
		π (bar)	0.0
		R (%)	N/A
Boundary flux data		Boundary flux type	C
		α (L h^{-2} m^{-2} bar^{-1})	—
		Δw% (%)	—
		J$_b$ (L h^{-1} m^{-2})	575.0
		TMP$_b$ (bar)	1.9

TABLE 7.92 Tomato Industry Wastewater

Reference ID		[25]
Feedstock	Key parameter	COD
	Value in feed stream	1200 mg L^{-1}
	Pretreatments	Biological pretreatment step
Membrane properties	Membrane type	NF
	Membrane model	Spiral wound TFC PA/PS
	Membrane ID	Desal-5 DK2540
	Membrane supplier	GE Osmonics
	Pore size	0.5 nm
	w (L h^{-1} m^{-2} bar^{-1})	7.9
Process properties	T (°C)	20.0
	F$_F$ (L h^{-1})	600.0
	π (bar)	0.001
	R (%)	60.9
Boundary flux data	Boundary flux type	C
	α (L h^{-2} m^{-2} bar^{-1})	–
	Δw% (%)	–
	J$_b$ (L h^{-1} m^{-2})	8.2
	TMP$_b$ (bar)	4.5

TABLE 7.93 Whey Solution

		Reference ID			
		[133]			
Feedstock	Key parameter	COD			
	Value in feed stream	100 g L⁻¹			
Membrane properties	Pretreatments	None	None	None	RO
	Membrane type	UF	NF	RO	NF
	Membrane model	Spiral wound PES	Spiral wound polysulfone	Spiral wound polyamide-urea	Spiral wound polyamide-urea
	Membrane ID	N/A	N/A	N/A	N/A
	Membrane supplier	Nadir GmbH	Trisep Corp.		
	Pore size	200 nm	20 nm	–	–
	w (L h⁻¹ m⁻² bar⁻¹)	430.0	250.0	190.0	190.0
Process properties	T (°C)	50.0			
	v_F (m s⁻¹)	2.0	2.5	3.0	3.0
	π (bar)	0.64		3.0	3.0
	R (%)	N/A	N/A	N/A	N/A
Boundary flux data	Boundary flux type	TH	TH	TH	TH
	α (L h⁻² m⁻² bar⁻¹)	3.630	0.540	0.029	0.025
	$\Delta w\%$ (%)	N/A	N/A	N/A	N/A
	J_b (L h⁻¹ m⁻²)	14.7	15.4	1.7	2.6
	TMP_b (bar)	3.0	8.0	12.0	12.0

TABLE 7.94 Xylene Solution

Reference ID		[134]
Feedstock	Key parameter	p-xylene concentration
	Value in feed stream	3000 mg L^{-1}
	Pretreatments	Powdered activated carbon
Membrane properties	Membrane type	Ceramic MF
	Membrane model	Zirconia (ZrO$_2$) on alumina porous support
	Membrane ID	–
	Membrane supplier	Nanjing Jiusi High-Tech Co.
	Pore size	0.54 nm
	w (L h^{-1} m^{-2} bar^{-1})	1344.0
Process properties	T (°C)	35.0
	v$_F$ (m s^{-1})	4.5
	π (bar)	0.0
	R (%)	99.2
Boundary flux data	Boundary flux type	TH
	α (L h^{-2} m^{-2} bar^{-1})	3.66
	Δw^o (%)	N/A
	J$_b$ (L h^{-1} m^{-2})	598.1
	TMP$_b$ (bar)	2.6

TABLE 7.95 Yeast Suspension

Reference ID		[135]	
Feedstock	Key parameter	Yeast concentration (*Saccharomyces cerevisiae*)	
	Value in feed stream	2.5×10^7 cells mL^{-1}	
	Pretreatments	None	Yeast cells as secondary membrane
Membrane properties	Membrane type	MF	
	Membrane model	Flat-sheet PES	
	Membrane ID	MicroPES® 4F	
	Membrane supplier	Membrane GmbH	
	Pore size	450 nm	
	w (L h^{-1} m^{-2} bar^{-1})	600.0	
Process properties	T (°C)	25.0	
	v_F (m s^{-1})	1.0	
	π (bar)	0.0	0.0
	R (%)	N/A	N/A
Boundary flux data	Boundary flux type	TH	TH
	α (L h^{-2} m^{-2} bar^{-1})	2500.0	2750.0
	Δw% (%)	N/A	N/A
	J_b (L h^{-1} m^{-2})	412.7	45.9
	TMP$_b$ (bar)	2.0	2.0

List of Variables

Variable	Description	Units
$	Total costs	\$, currency
$c'	Membrane investment costs	\$, currency
$c''	Pump investment costs	\$, currency
$cp	Cost of one pump	\$, currency
$h	Cost of one membrane housing	\$, currency
$m	Cost of one membrane module	\$, currency
$o	Pump operating costs	\$, currency
a	Fitting parameter of the sub-threshold flux equation	$l\ h^{-2}\ m^{-1}\ bar^{-1}$
A	Required membrane area	m^2
A_m	Membrane area of one membrane module	m^2
A_p	Feed stream passage area of the membrane module	m^2
B	Fitting parameter of the critical flux equation	$h^{-1}\ m^{-2}\ bar^{-1}$
b	Fitting parameter of the super-threshold flux equation	$h^{-1}\ m^{-2}\ bar^{-1}$
C	Number of expected separation cycles of the membranes	—
F_F	Feed stream flow rate	$l\ h^{-1}$
F_p	Permeate flow rate	$l\ h^{-1}$
F_R	Required project flow rate	$l\ h^{-1}$
J_c	Critical flux	$l\ h^{-1}\ m^{-2}$
J_p	Permeate flux	$l\ h^{-1}\ m^{-2}$
J_{th}	Threshold flux	$l\ h^{-1}\ m^{-2}$
KP	Key parameter value of the feedstock	[KP], variable
m	Membrane permeability	$l\ h^{-1}\ m^{-2}\ bar^{-1}$
M	Number of membrane modules per each membrane housing	—
m'	Fitting parameter of the permeability equation	$l\ h^{-1}\ m^{-2}\ bar^{-1}\ [KP]^{-1}$
N	Number of membrane modules	—
N_p	Number of pumps	—
N_M	Number of membrane module parallel lines	—
N_S	Number of membrane modules in series in each line	—
p	Fitting parameter of the osmotic pressure equation	$bar\ [KP]^{-1}$
R	Rejection	—
t	Time	h

(Continued)

cont'd

Variable	Description	Units
T	Temperature	°C
TMP	Transmembrane pressure	bar
TMP_b	Boundary transmembrane pressure	bar
V	Volume	l
v_F	Feed stream velocity	$m\ h^{-1}$
w	Pure water permeability	$l\ h^{-1}\ m^{-2}\ bar^{-1}$
Y	Recovery factor	—
α	Sub-boundary fouling index	$l\ h^{-2}\ m^{-1}\ bar^{-1}$
β	Super-boundary fouling index	$h^{-1}\ bar^{-1}$
γ	Fitting parameter of the rejection equation	—
δ	Design safety value	—
Δw%	Cleaning efficiency	h^{-1}
ε	Fitting parameter of the super-boundary fouling index equation	—
ζ	Fitting parameter of the super-boundary fouling index equation	$h^{-1}\ bar^{-2}$
π	Osmotic pressure	bar
ρ	feed stream density	$kg\ L^{-1}$
σ	Reflection coefficient	—
τ	Period of time between cleaning cycles	h
$τ_P$	Expected pump lifetime	h
φ	Expected plant service time	h

References

[1] Ho WSW, Sirkar KK. Membrane handbook. Chapman & Hall; 1992.

[2] Saad MA. Membrane desalination for the Arab world: overview and outlook. Arab Water World 2005;1:10−4.

[3] Field RW, Wu D, Howell JA, Gupta BB. Critical flux concept for microfiltration fouling. J Memb Sci 1995;100:259−72.

[4] Mänttäri M, Nyström M. Critical flux in NF of high molar mass polysaccharides and effluents from the paper industry. J Memb Sci 2000;170:257−73. http://dx.doi.org/10.1016/S0376-7388(99)00373-7.

[5] Bacchin P, Aimar P, Field RW. Critical and sustainable fluxes: theory, experiments and applications. J Memb Sci 2006;281:42−69.

[6] http://www.waterindustry.org/New%20Projects/membrane-3.htm; 2013.

[7] Aachener Membran Kolloquium. Aachen, Germany: VDI; 2001, ISBN 3-89653-834-9.

[8] Hassan M, Jamaluddin T, Saeed O, Al-Rubaian A, Al-Reweli A. Al-Birk "SWRO plant operation with Toyobo membrane in train 200 instead of Dupont b-10 membrane", SWCC Research Paper.

[9] Pikorová T. Two years of the operation of a domestic MBR wastewater treatment plant. Slovak J Civ Eng 2006;VXX(2):28−36.

[10] Le-Clech P, Chen V, Fane TAG. Fouling in membrane bioreactors used in wastewater treatment. J Memb Sci 2006;284(1−2):17−53.

[11] Field RW, Pearce GK. Critical, sustainable and threshold fluxes for membrane filtration with water industry applications. Adv Colloid Interface Sci 2011;164(1−2):38−44.

[12] Stoller M, Chianese A. Optimization of membrane batch processes by means of the critical flux theory. Desalination 2006;191:62−70.

[13] Stoller M, Chianese A. Influence of the adopted pretreatment process on the critical flux value of batch membrane processes. Ind Eng Chem Res 2007;8(46):2249−53.

[14] Lipp P, Lee CH, Fane AG, Fell CJD. A fundamental study of the UF of oil−water emulsions. J Memb Sci 1988;36:161−77.

[15] Data taken from Scopus. December 2013. http://www.scopus.com.

[16] Cicci A, Stoller M, Bravi M. Microalgal biomass production by using ultra- and nanofiltration membrane fractions of olive mill waste water. Water Res 2013;47:4710−8. http://dx.doi.org/10.1016/j.watres.2013.05.030.

[17] Ochando-Pulido JM, Stoller M, Bravi M, Martinez-Ferez A, Chianese A. Batch membrane treatment of olive vegetation wastewater from two-phase olive oil production process by threshold flux based methods. Sep Purif Technol 2012;101:34−41. http://dx.doi.org/10.1016/j.seppur.2012.09.015.

[18] Stoller M, De Caprariis B, Cicci A, Verdone N, Bravi M, Chianese A. About proper membrane process design affected by fouling by means of the analysis of measured threshold flux data. Sep Purif Technol 2013;114:83−9. http://dx.doi.org/10.1016/j.seppur.2013.04.041.

[19] Stoller M, Ochando Pulido JM, Chianese A. Comparison of critical and threshold fluxes on ultrafiltration and nanofiltration by treating 2-phase or 3-phase olive mill wastewater. Chem Eng Trans 2013;32:397−402. http://dx.doi.org/10.3303/CET1332067.

[20] Stoller M. A three year long experience of effective fouling inhibition by threshold flux based optimization methods on a NF membrane module for olive mill wastewater treatment. Chem Eng Trans 2013;32:37−42. http://dx.doi.org/10.3303/CET1332007.

[21] Luo J, Zhu Z, Ding L, Bals O, Wan Y, Jaffrin MY, et al. Flux behavior in clarification of chicory juice by high-shear membrane filtration: evidence for threshold flux. J Memb Sci 2013;435:120−9.

[22] Barredo-Damas S, Alcaina-Miranda MI, Iborra-Clar MI, Mendoza-Roca JA. Application of tubular ceramic ultrafiltration membranes for the treatment of integrated textile wastewaters. Chem Eng J 2012;192:211−8.

[23] Wicaksana F, Fane AG, Pongpairoj P, Field R. Microfiltration of algae (*Chlorella sorokiniana*): critical flux, fouling and transmission. J Memb Sci 2012;387−388:83−92. http://dx.doi.org/10.1016/j.memsci.2011.10.013.

[24] Stoller M. Technical optimization of a dual ultrafiltration and nanofiltration pilot plant in batch operation by means of the critical flux theory: a case study. Chem Eng Process J 2008;47(7):1165−70.

[25] Iaquinta M, Stoller M, Merli C. Optimization of a nanofiltration membrane process for tomato industry wastewater effluent treatment. Desalination 2009;245:314−20. http://dx.doi.org/10.1016/j.desal.2008.05.028.

[26] Stoller M, Di Palma L, Merli C. Optimisation of batch membrane processes for the removal of residual heavy metal contamination in pretreated marine sediment. Chem Ecol 2011;27:171−9. http://dx.doi.org/10.1080/02757540.2010.534083.

[27] Stoller M, Chianese A. Technical optimization of a batch olive wash wastewater treatment membrane plant. Desalination 2006;200:734−6.

[28] Stoller M. Effective fouling inhibition by critical flux based optimization methods on a NF membrane module for olive mill wastewater treatment. Chem Eng J 2011;168(3):1140−8.

[29] Stoller M, Bravi M. Critical flux analyses on differently pretreated olive vegetation waste water streams: some case studies. Desalination 2010;250(2):578−82.

[30] Lim AL, Rembi B. Membrane fouling and cleaning in MF of activated sludge wastewater. J Memb Sci 2003;216:279−90.

[31] Stoller M. On the effect of flocculation as pre-treatment process for membrane fouling reduction. Desalination 2009;240:209−17.

[32] Aimar P, Bacchin P. Slow colloidal aggregation and membrane fouling. J Memb Sci 2010;360:70−6.

[33] Ognier S, Wisniewski C, Grasmick A. Membrane bioreactor fouling in sub-critical filtration conditions: a local critical flux concept. J Memb Sci 2004;229:171−7.

[34] Russo C. A new membrane process for the selective fractionation and total recovery of polyphenols, water and organic substances from vegetation waters. J Memb Sci 2007;288(1−2):239−46.

[35] Tsagaraki EV, Lazarides HN. Fouling analysis and performance of tubular ultrafiltration on pretreated olive mill waste water. Food Bioprocess Technol 2012;5(2):584−92.

[36] Zhou M, Liu H, Kilduff JE, Langer R, Anderson DG, Belfort G. High-throughput membrane surface modification to control NOM fouling. Environ Sci Technol 2009;43(10):3865−71.

[37] Dhaouadi H, Marrot B. Olive mill wastewater treatment in a membrane bioreactor: process stability and fouling aspects. Environ Technol 2010;31(7):761−70.

[38] El-Abbassi A, Hafidi A, Khayet M, García-Payo MC. Integrated direct contact membrane distillation for olive mill wastewater treatment. Desalination 2013;323:31–8.

[39] Stoller M, Sacco O, Sannino D, Chianese A. Successful integration of membrane technologies in a conventional purification process of tannery wastewater streams. Membranes 2013;3:126–35. http://dx.doi.org/10.3390/membranes3030126.

[40] Sacco O, Stoller M, Vaiano V, Ciambelli P, Chianese A, Sannino D. Photocatalytic degradation of organic dyes under visible light on N-doped photocatalysts. Int J Photoenergy 2012;2012. http://dx.doi.org/10.1155/2012/626759. Article ID 626759, 8 pages.

[41] Stoller M, Movassaghi K, Chianese A. Photocatalytic degradation of orange II in aqueous solution by immobilized nanostructured titanium dioxide. Chem Eng Trans 2011;24: 229–34.

[42] Sannino D, Vaiano V, Isupova LA, Ciambelli P. Heterogeneous photo-Fenton oxidation of organic pollutants on structured catalysts. J Adv Oxid Technol 2012;15(2):294–300.

[43] Di Palma L, Merli C, Petrucci E. Oxidation of phosphorous compounds by Fenton reagent. Ann Chim – Roma 2003;93(11):935–43.

[44] Di Palma L, Gonzini O, Mecozzi R. Effect of acidification and modified Fenton treatment on contaminated marine harbour sediments. Chem Ecol 2011;27:153–60. http://dx.doi.org/10.1080/02757540.2011.534237.

[45] Hodaifa G, Ochando-Pulido JM, Rodriguez-Vives S, Martinez-Ferez A. Optimization of continuous reactor at pilot scale for olive-oil mill wastewater treatment by Fenton-like process. Chem Eng J 2013;220:117–24.

[46] Ruzmanova Y, Stoller M, Chianese A. Photocatalytic treatment of olive mill waste water by magnetic core titanium dioxide nanoparticles. Chem Eng Trans 2013;32:2269–74. http://dx.doi.org/10.3303/CET1332379.

[47] Ruzmanova Y, Ustundas M, Stoller M, Chianese A. Photocatalytic treatment of olive mill waste water by N-doped titanium dioxide nanoparticles under visible light. Chem Eng Trans 2013;32:2233–8. http://dx.doi.org/10.3303/CET1332373.

[48] Stoller M, Ochando-Pulido JM. About merging threshold and critical flux concepts into a single one: the boundary flux. Sci World J 2014. http://dx.doi.org/10.1155/2014/656101. Article ID 656101, 8 pages.

[49] De Caprariis B, Di Rita M, Stoller M, Verdone N, Chianese A. Reaction-precipitation by a spinning disc reactor: influence of hydrodynamics on nanoparticles production. Chem Eng Sci 2012;76:73–80.

[50] Parisi M, Stoller M, Chianese A. Production of nanoparticles of hydroxyapatite by using a rotating disk reactor. Chem Eng Trans 2011;24:211–6.

[51] Stoller M, Miranda L, Chianese A. Optimal feed location in a spinning disc reactor for the production of TiO_2 nanoparticles. Chem Eng Trans 2009;17:993–8.

[52] Espinasse B, Bacchin P, Aimar P. On an experimental method to measure critical flux in ultrafiltration. Desalination 2002;146:91–6.

[53] Vyas HK, Bennett RJ, Marshall AD. Performance of cross flow MF during constant TMP and constant flux operations. Int Diary J 2002;12:473–9.

[54] McCool BC, Rahardianto A, Faria J, Kovac K, Lara D, Cohen Y. Feasibility of reverse osmosis desalination of brackish agricultural drainage water in the San Joaquin Valley. Desalination 2008;261:240–50. http://dx.doi.org/10.1016/j.desal.2010.05.031.

[55] Caus A, Vanderhaegen S, Braeken L, Van der Bruggen B. Integrated nanofiltration cascades with low salt rejection for complete removal of pesticides in drinking water production. Desalination 2009;241:111–7. http://dx.doi.org/10.1016/j.desal.2008.01.061.

[56] Zhang X, Fan L, Roddick FA. Influence of the characteristics of soluble algal organic matter released from *Microcystis aeruginosa* on the fouling of a ceramic microfiltration membrane. J Memb Sci 2013;425−426:23−9. http://dx.doi.org/10.1016/j.memsci.2012.09. 033.

[57] Cancino-Madariaga B, Hurtado CF, Ruby R. Effect of pressure and pH in ammonium retention for nanofiltration and reverse osmosis membranes to be used in recirculation aquaculture systems (RAS). Aquacult Eng 2011;45:103−8. http://dx.doi.org/10.1016/j. aquaeng.2011.08.002.

[58] Dagnew M, Parker W, Seto P. Anaerobic membrane bioreactors for treating waste activated sludge: short term membrane fouling characterization and control tests. J Memb Sci 2012;421−422:103−10. http://dx.doi.org/10.1016/j.memsci.2012.06.046.

[59] Tanhaei B, Chenar MP, Saghato-leslami N, Hesampour M, Laakso T, Kallioinen M, et al. Simultaneous removal of aniline and nickel from water by micellar-enhanced ultrafiltration with different molecular weight cut-off membranes. Sep Purif Technol 2014;124:26−35. http://dx.doi.org/10.1016/j.seppur.2014.01.009.

[60] Kertész S, László Z, Horváth ZH, Hodúr C. Analysis of nanofiltration parameters of removal of an anionic detergent. Desalination 2008;221:303−11. http://dx.doi.org/10. 1016/j.desal.2007.01.087.

[61] Teychene B, Collet G, Gallard H, Croue J. A comparative study of boron and arsenic (III) rejection from brackish water by reverse osmosis membranes. Desalination 2013;310:109−14. http://dx.doi.org/10.1016/j.desal.2012.05.034.

[62] Madsen HT, Søgaard EG. Applicability and modelling of nanofiltration and reverse osmosis for remediation of groundwater polluted with pesticides and pesticide transformation products. Sep Purif Technol 2014;125:111−9. http://dx.doi.org/10.1016/j. seppur.2014.01.038.

[63] Metsämuuronen S, Howell J, Nyström M. Critical flux in ultrafiltration of myoglobin and baker's yeast. J Memb Sci 2002;196:13−25. http://dx.doi.org/10.1016/S0376-7388(01) 00572-5.

[64] Alzahrani S, Mohammad AW, Abdullah P, Jaafar O. Potential tertiary treatment of produced water using highly hydrophilic nanofiltration and reverse osmosis membranes. J Environ Chem Eng 2013;1:1341−9. http://dx.doi.org/10.1016/j.jece.2013.10.002.

[65] Agana BA, Reeve D, Orbell JD. Performance optimization of a 5 nm TiO_2 ceramic membrane with respect − beverage production wastewater. Desalination 2013;311:162−72. http://dx.doi.org/10.1016/j.desal.2012.11.027.

[66] Tu KL, Chivas AR, Nghiem LD. Effects of membrane fouling and scaling on boron rejection by nanofiltration and reverse osmosis membranes. Desalination 2011;279:269−77. http://dx. doi.org/10.1016/j.desal.2011.06.019.

[67] Koseoglu H, Kabay N, Yüksel M, Kitis M. The removal of boron from model solutions and seawater using reverse osmosis membranes. Desalination 2008;223:126−33. http://dx.doi. org/10.1016/j.desal.2007.01.189.

[68] Yao M, Zhang K, Cui L. Characterization of protein−polysaccharide ratios on membrane fouling. Desalination 2010;259:11−6. http://dx.doi.org/10.1016/j.desal.20.10.04.049.

[69] Zhao L, Chang PCY, Yen C, Winston Ho WS. High-flux and fouling-resistant membranes for brackish water desalination. J Memb Sci 2013;425−426:1−10. http://dx.doi.org/ 10.1016/j.memsci.2012.09.018.

[70] Zaky A, Escobar I, Motlagh AM, Gruden C. Determining the influence of active cells and conditioning layer on early stage biofilm formation using cellulose acetate ultrafiltration membranes. Desalination 2012;286:296−303. http://dx.doi.org/10.1016/j.desal.2011.11.040.

[71]   Häyrynen P, Landaburu-Aguirre J, Pongrácz E, Keiski RL. Study of permeate flux in micellar-enhanced ultrafiltration on a semi-pilot scale: simultaneous removal of heavy metals from phosphorous rich real wastewaters. Sep Purif Technol 2012;93:59−66. http://dx.doi.org/10.1016/j.seppur.2012.03.029.

[72]   Agana BA, Reeve D, Orbell JD. Optimization of the operational parameters for a 50 nm ZrO_2 ceramic membrane as applied − the ultrafiltration of post-electrodeposition rinse wastewater. Desalination 2012;278:325−32. http://dx.doi.org/10.1016/j.desal.2011.05.043.

[73]   Bade R, Lee SH, Jo S, Lee H-S, Lee S. Micellar enhanced ultrafiltration (MEUF) and activated carbon fibre (ACF) hybrid processes for chromate removal from wastewater. Desalination 2007;229:264−78. http://dx.doi.org/10.1016/j.desal.2007.10.015.

[74]   Daniş Ü, Keskinler B. Chromate removal from wastewater using micellar enhanced crossflow filtration: effect of transmembrane pressure and crossflow velocity. Desalination 2009;249:1356−64. http://dx.doi.org/10.1016/j.desal.2009.06.023.

[75]   Capar G. Separation of silkworm proteins in cocoon cooking wastewaters via nano-filtration: effect of solution pH on enrichment of sericin. J Memb Sci 2012;389:509−21. http://dx.doi.org/10.1016/j.memsci.2011.11.021.

[76]   Yin N, Yang G, Zhong Z, Xing W. Separation of ammonium salts from coking wastewater with nanofiltration combined with diafiltration. Desalination 2011;268:233−7. http://dx. doi.org/10.1016/j.desal.20.10.034.

[77]   Popović S, Milanović S, Iličić M, Djurić M, Tekić M. Flux recovery of tubular ceramic membranes fouled with whey proteins. Desalination 2009;249:293−300. http://dx.doi.org/ 10.1016/j.desal.2008.12.060.

[78]   Rice GS, Kentish SE, O'Connor AJ, Barber AR, Pihlajamäki A, Nyström M, et al. Analysis of separation and fouling behaviour during nanofiltration of dairy ultrafiltration permeates. Desalination 2009;236:23−9. http://dx.doi.org/10.1016/j.desal.2007.10.046.

[79]   Luo J, Ding L. Influence of pH on treatment of dairy wastewater by nanofiltration using shear-enhanced filtration system. Desalination 2011;278:150−6. http://dx.doi.org/10.1016/ j.desal.2011.05.025.

[80]   Huang J, Zhang K. The high flux poly (m-phenylene isophthalamide) nanofiltration membrane for dye purification and desalination. Desalination 2011;208:19−26. http://dx. doi.org/10.1016/j.desal.2011.09.045.

[81]   Al-Juboori RA, Yusaf T, Aravinthan V. Investigating the efficiency of thermosonication for controlling biofouling in batch membrane systems. Desalination 2012;286:349−57. http://dx.doi.org/10.1016/j.desal.2011.11.049.

[82]   Chiou Y, Hsieh M, Yeh H. Effect of algal extracellular polymer substances on UF membrane fouling. Desalination 2010;250:648−52. http://dx.doi.org/10.1016/j.desal.2008.02.043.

[83]   Wang Z, Song Y, Liu M, Yao J, Wang Y, Hu Z, et al. Experimental study of filterability behavior of model extracellular polymeric substance solutions in dead-end membrane filtration. Desalination 2009;249:1380−4. http://dx.doi.org/10.1016/j.desal.2008.06.028.

[84]   Singh AK, Thakur AK, Shahi VK. Self-assembled nanofiltration membrane containing antimicrobial organosilica prepared by sol−gel process. Desalination 2013;309:275−83. http://dx.doi.org/10.1016/j.desal.2012.10.011.

[85]   Chakrabortty S, Roy M, Pal P. Removal of fluoride from contaminated groundwater by cross flow nanofiltration: transport modeling and economic evaluation. Desalination 2013;313:115−24. http://dx.doi.org/10.1016/j.desal.2012.12.021.

[86]   Abbasi Monfared M, Kasiri N, Salahi A, Mohammadi T. CFD simulation of baffles arrangement for gelatin-water ultrafiltration in rectangular channel. Desalination 2012;284:288−96. http://dx.doi.org/10.1016/j.desal.2011.09.014.

[87] Yavuz E, Güler E, Sert G, Arar Ö, Yüksel M, Yüksel Ü, et al. Removal of boron from geothermal water by RO system-I—effect of membrane configuration and applied pressure. Desalination 2013;310:130—4. http://dx.doi.org/10.1016/j.desal.2012.07.026.

[88] Walha K, Ben Amar R, Quemeneur F, Jaouen P. Treatment by nanofiltration and reverse osmosis of high salinity drilling water for seafood washing and processing abstract. Desalination 2008;219:231—9. http://dx.doi.org/10.1016/j.desal.2007.05.016.

[89] Abu Seman MN, Khayet M, Hilal N. Comparison of two different UV-grafted nanofiltration membranes prepared for reduction of humic acid fouling using acrylic acid and N-vinylpyrrolidone. Desalination 2012;287:19—29. http://dx.doi.org/10.1016/j.desal.20.10.031.

[90] Soffer Y, Adin A, Gilron J. Threshold flux in fouling of OF membranes by colloidal iron Desalination 2004;161(3):207—21. http://dx.doi.org/10.1016/S0011-9164(03)00702-1.

[91] Sikder J, Pereira C, Palchoudhury S, Vohra K, Basumatary D, Pal P. Synthesis and characterization of cellulose acetate-polysulfone blend microfiltration membrane for separation of microbial cells from lactic acid fermentation broth. Desalination 2009;249:802—8. http://dx.doi.org/10.1016/j.desal.2008.11.024.

[92] Mariam T, Nghiem LD. Landfill leachate treatment using hybrid coagulation-nanofiltration processes. Desalination 2010;250:677—81. http://dx.doi.org/10.1016/j.desal.2009.03.024.

[93] Ince M, Senturk E, Onkal Engin G, Keskinler B. Further treatment of landfill leachate by nanofiltration and microfiltration—PAC hybrid process. Desalination 2010;255:52—60. http://dx.doi.org/10.1016/j.desal.20.10.01.017.

[94] Gherasim C, Mikulášek P. Influence of operating variables on the removal of heavy metal ions from aqueous solutions by nanofiltration. Desalination 2014;343:67—74. http://dx.doi.org/10.1016/j.desal.2013.11.012.

[95] Kazemi MA, Soltanieh M, Yazdanshenas M. Modeling of transient permeate flux decline during crossflow microfiltration of non-alcoholic beer with consideration of particle size distribution. J Memb Sci 2012;411—412:13—21. http://dx.doi.org/10.1016/j.memsci.2012.02.064.

[96] Ahmad AL, Mat Yasin NH, Derek CJC, Lim JK. Crossflow microfiltration of microalgae biomass for biofuel production. Desalination 2012;302:65—70. http://dx.doi.org/10.1016/j.desal.2012.06.026.

[97] Rickman M, Pellegrino J, Davis R. Fouling phenomena during membrane filtration of microalgae. J Memb Sci 2012;423—424:33—42. http://dx.doi.org/10.1016/j.memsci.2012.07.013.

[98] Frappart M, Massé A, Jaffrin MY, Pruvost J, Jaouen P. Influence of hydrodynamics in tangential and dynamic ultrafiltration systems for microalgae separation. Desalination 2011;265:279—83. http://dx.doi.org/10.1016/j.memsci.2008.01.028.

[99] Veerasamy D, Supurmaniam A, Nor ZM. Evaluating the use of in-situ ultrasonication-reduce fouling during natural rubber skim latex (waste latex) recovery by ultrafiltration. Desalination 2009;236:202—7. http://dx.doi.org/10.1016/j.desal.2007.10.068.

[100] Mattaraj S, Jarusutthirak C, Charoensuk C, Jiraratananon R. A combined pore blockage, osmotic pressure, and cake filtration model for crossflow nanofiltration of natural organic matter and inorganic salts. Desalination 2011;274:182—91. http://dx.doi.org/10.1016/j.desal.2011.02.010.

[101] Santafé-Moros A, Gozálvez-Zafrilla JM, Lora-García J. Performance of commercial nanofiltration membranes in the removal of nitrate ions. Desalination 2005;185:281—7. http://dx.doi.org/10.1016/j.desal.2005.02.080.

[102]   Falahati H, Tremblay AY. Flux dependent oil permeation in the ultrafiltration of highly concentrated and unstable oil-in-water emulsions. J Memb Sci 2011;371:239−47. http://dx.doi.org/10.1016/j.memsci.2011.01.047.

[103]   Wandera D, Husson SM. Assessment of fouling-resistant membranes for additive-free treatment of high-strength wastewaters. Desalination 2013;309:222−30. http://dx.doi.org/10.1016/j.desal.2012.10.013.

[104]   Mittal P, Jana S, Mohanty K. Synthesis of low-cost hydrophilic ceramic−polymeric composite membrane for treatment of oily wastewater. Desalination 2011;282:54−62. http://dx.doi.org/10.1016/j.desal.2011.06.071.

[105]   Ochando-Pulido JM, Rodriguez-Vives S, Martinez-Ferez A. The effect of permeate recirculation on the depuration of pretreated olive mill wastewater through reverse osmosis membranes. Desalination 2012;286:145−54. http://dx.doi.org/10.1016/j.desal.2011.10.041.

[106]   Stoller M, Bravi M, Chianese A. Threshold flux measurements of a nanofiltration membrane module by critical flux data conversion. Desalination 2013;315:142−8. http://dx.doi.org/10.1016/j.desal.2012.11.013.

[107]   Kochkodan V, Tsarenko S, Potapchenko N, Kosinova V, Goncharuk V. Adhesion of microorganisms-polymer membranes: a pho-bactericidal effect of surface treatment with TiO_2. Desalination 2008;220:380−5. http://dx.doi.org/10.1016/j.desal.2007.01.042.

[108]   Vincent Vela MC, Bergantiños-Rodríguez E, Álvarez Blanco S, García JL. Influence of feed concentration on the accuracy of permeate flux decline prediction in ultrafiltration. Desalination 2008;221:383−9. http://dx.doi.org/10.1016/j.desal.2007.01.097.

[109]   Madaeni SS, Monfared HA, Vatanpour V, Shamsabadi AA, Salehi E, Daraei P, et al. Coke removal from petrochemical oily wastewater using γ-Al_2O_3 based ceramic microfiltration membrane. Desalination 2012;293:87−93. http://dx.doi.org/10.1016/j.desal.2012.02.028.

[110]   Mosqueda-Jimenez DB, Huck PM. Effect of biofiltration as pretreatment on the fouling of nanofiltration membranes. Desalination 2009;245:60−72. http://dx.doi.org/10.1016/j.desal.2008.05.027.

[111]   Inchaurrondo N, Haure P, Font J. Nanofiltration of partial oxidation products and copper from catalyzed wet peroxidation of phenol. Desalination 2013;315:76−82. http://dx.doi.org/10.1016/j.desal.2012.12.024.

[112]   Arkhangelsky E, Steubing B, Ben-Dov E, Kushmaro A, Gitis V. Influence of pH and ionic strength on transmission of plasmid DNA through ultrafiltration membranes. Desalination 2008;227:111−9. http://dx.doi.org/10.1016/j.desal.2007.07.017.

[113]   Zhang H, Zhong Z, Xing W. Application of ceramic membranes in the treatment of oilfield-produced water: effects of polyacrylamide and inorganic salts. Desalination 2013;309:84−90. http://dx.doi.org/10.1016/j.desal.2012.09.012.

[114]   Fang J, Qin G, Wei W, Zhao X, Jiang L. Elaboration of new ceramic membrane from spherical fly ash for microfiltration of rigid particle suspension and oil-in-water emulsion. Desalination 2013;311:113−26. http://dx.doi.org/10.1016/j.desal.2012.11.008.

[115]   Hwang K, Liao C, Tung K. Effect of membrane pore size on the particle fouling in membrane filtration. Desalination 2008;234:16−23. http://dx.doi.org/10.1016/j.desal.2007.09.065.

[116]   Han MJ, Nathaniel G, Baroña B, Jung B. Effect of surface charge on hydrophilically modified poly(vinylidene fluoride) membrane for microfiltration. Desalination 2011;270:76−83. http://dx.doi.org/10.1016/j.desal.20.10.11.024.

[117]   Yu C, Wu J, Contreras AE, Li Q. Control of nanofiltration membrane biofouling by *Pseudomonas aeruginosa* using D-tyrosine. J Memb Sci 2012;423−424:487−94. http://dx.doi.org/10.1016/j.memsci.2012.08.051.

[118]   Li M, Zhao Y, Zhou S, Xing W. Clarification of raw rice wine by ceramic microfiltration membranes and membrane fouling analysis. Desalination 2010;256:166−73. http://dx.doi. org/10.1016/j.desal.20.10.01.018.

[119]   Pezeshk N, Narbaitz RM. More fouling resistant modified PVDF ultrafiltration membranes for water treatment. Desalination 2012;287:247−54. http://dx.doi.org/10.1016/j.desal. 2011.11.048.

[120]   El Rayess Y, Albasi C, Bacchin P, Taillandier P, Miet-n-Peuchot M, Devatine A. Cross-flow microfiltration of wine: effect of colloids on critical fouling conditions. J Memb Sci 2011;385−386:177−86. http://dx.doi.org/10.1016/j.memsci.2011.09.037.

[121]   Zaghbani N, Hafiane A, Dhahbi M. Removal of Safranin T from wastewater using micellar enhanced ultrafiltration. Desalination 2008;222:348−56. http://dx.doi.org/10.1016/j.desal. 2007.01.148.

[122]   Jeong S, Choi YJ, Nguyen TV, Vigneswaran S, Hwang TM. Submerged membrane hybrid systems as pretreatment in seawater reverse osmosis (SWRO): optimisation and fouling mechanism determination. J Memb Sci 2012;411−412:173−81. http://dx.doi.org/10.1016/ j.memsci.2012.04.029.

[123]   Xiao Y, Liu XD, Wang DX, Lin YK, Han YP, Wang XL. Feasibility of using an innovative PVDF MF membrane prior-RO for reuse of a secondary municipal effluent. Desalination 2013;311:16−23. http://dx.doi.org/10.1016/j.desal.2012.10.022.

[124]   Jiang Q, Rentschler J, Perrone R, Liu K. Application of ceramic membrane and ion-exchange for the treatment of the flowback water from Marcellus shale gas production. J Memb Sci 2013;431:55−61. http://dx.doi.org/10.1016/j.memsci.2012.12.030.

[125]   Zhang Y, Xu Y, Zhang S, Zhang Y, Xu Z. Study on a novel composite membrane for treatment of sewage containing oil. Desalination 2012;299:63−9. http://dx.doi.org/10. 1016/j.desal.2012.05.020.

[126]   Huisman IH, Vellenga E, Trägårdh G, Trägårdh C. The influence of the membrane zeta potential on the critical flux for crossflow microfiltration of particle suspensions. J Memb Sci 1999;156:153−8. http://dx.doi.org/10.1016/S0376-7388(98)00328-7.

[127]   Musbah I, Cicéron D, Saboni A, Alexandrova S. Retention of pesticides and metabolites by nanofiltration by effects of size and dipole moment. Desalination 2013;313:51−6. http://dx.doi.org/10.1016/j.desal.2012.11.016.

[128]   Li F, Tian Q, Yang B, Wu L, Deng C. Effect of polyvinyl alcohol addition-model extra-cellular polymeric substances (EPS) on membrane filtration performance. Desalination 2012;286:34−40. http://dx.doi.org/10.1016/j.desal.2011.10.035.

[129]   Sérgi Gomes MC, Arroyo PA, Curvelo Pereira N. Influence of acidified water addition on the biodiesel and glycerol separation through membrane technology. J Memb Sci 2013;431:28−36. http://dx.doi.org/10.1016/j.memsci.2012.12.036.

[130]   Abbasi M, Salahi A, Mirfendereski M, Mohammadi T, Pak A. Dimensional analysis of permeation flux for microfiltration of oily wastewaters using mullite ceramic membranes. Desalination 2010;252:113−9. http://dx.doi.org/10.1016/j.desal.2009.10.015.

[131]   Ellouze E, Tahri N, Amar RB. Enhancement of textile wastewater treatment process using nanofiltration. Desalination 2012;286:16−23. http://dx.doi.org/10.1016/j.desal.2011. 09.025.

[132]   Chiu T, James A. Critical flux determination of non-circular multi-channel ceramic membranes using TiO suspensions. J Memb Sci 2005;254:295−301. http://dx.doi.org/10. 1016/j.memsci.2005.01.037.

[133] Yorgun MS, Akmehmet Balcioglu I, Saygin O. Performance comparison of ultrafiltration, nanofiltration and reverse osmosis on whey treatment. Desalination 2008;229:204−16. http://dx.doi.org/10.1016/j.desal.2007.09.008.

[134] Yang Y, Chen R, Xing W. Integration of ceramic membrane microfiltration with powdered activated carbon for advanced treatment of oil-in-water emulsion. Sep Purif Technol 2011;76:373−7. http://dx.doi.org/10.1016/j.seppur.20.10.11.008.

[135] Chupakhina S, Kottke V. Using a yeast cell layer as a secondary membrane in microfiltration. Desalination 2008;224:18−22. http://dx.doi.org/10.1016/j.desal.2007. 04.073.

Index

Printed in the United States
By Bookmasters